I0491189

HISTORY OF MATHEMATICS: MESOPOTAMIA (ANCIENT IRAQ)

Saad Taha Bakir, Ph.D.

HISTORY OF MATHEMATICS: MESOPOTAMIA (ANCIENT IRAQ)

Saad Taha Bakir, Ph.D.

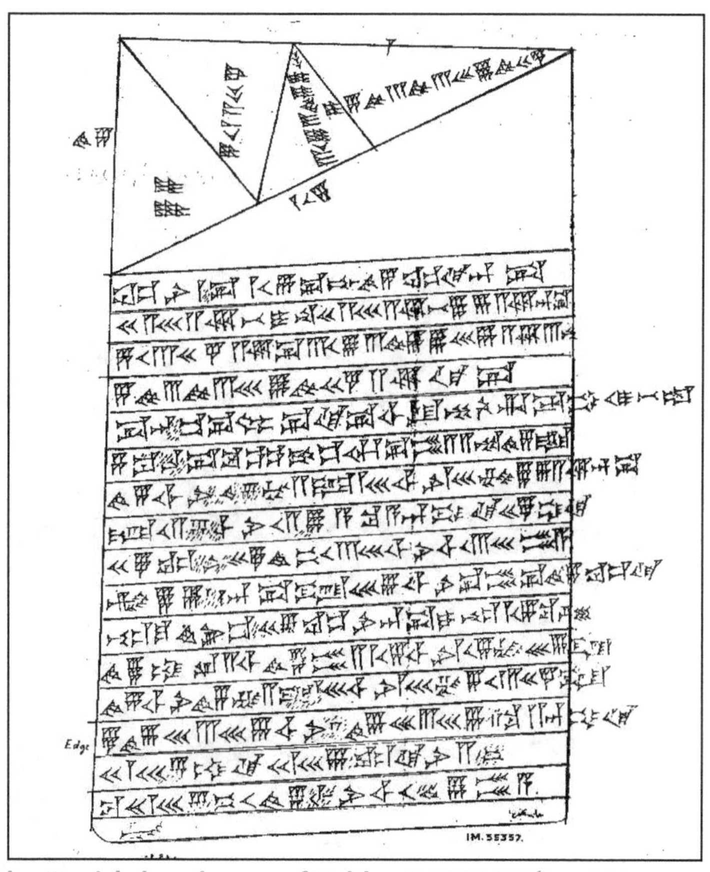

Taha Baqir's hand copy of Tablet IM 55357 (Iraqi Museum)

ISBN: 1974502619
ISBN 13: 9781974502615
Library of Congress Control Number: 2017916389
CreateSpace Independent Publishing Platform, North Charleston, SC

Cover Design: CreateSpace Cover Creator
Front cover: Mesopotamian tablet IM 55357 (1800 BCE) on similar triangles (Iraqi Museum)

Contents

Preface

This book presents a brief history of ancient mathematics that had developed in Mesopotamia (ancient Iraq), from the early Sumerian civilization (ca. 3500 BCE) up to the fall of the Seleucid rule (125 BCE).

I have written this book in tribute to the memory of my father, Taha Baqir (1912–84), the renowned Iraqi archeologist. Since the mid-1940s, Baqir had excavated, deciphered, and published dozens of mathematical and legal cuneiform tablets. In particular, he discovered the four-thousand-year-old laws of Eshnunna and translated the epic of Gilgamesh directly from Akkadian to Arabic. Baqir was proficient in the ancient Akkadian, Aramaic, and Sumerian languages as well as in modern Arabic, English, French, and German.

One vivid childhood memory of my father motivated me to write this book. Sometime in the late 1950s, my father noticed me reviewing mathematics for my high-school baccalaureate exam. He went away for a while and came back holding a copy of cuneiform tablet IM 55357 (see, Chapter 8) that he had excavated at Tell Harmal near Baghdad. He also handed me his English translation of the tablet. "When you have time, I would like you to go over the mathematical steps in this cuneiform tablet and let me know," he said and left the room. After reading the translation, I went back to him that afternoon and said, "Dad, the algebraic steps are all correct." I hoped that my answer would satisfy him, but it did not. "Great, but tell me what famous mathematical principles did these ancient people apply in their solution?" I had no answer at the time. "Are they not using the properties of similar triangles? Go get your geometry textbook," he asked. I did just that and handed him the geometry book. He showed me Euclid's Proposition 8, Book VI, stating: "If in a right-angled triangle, a perpendicular is drawn from the right angle to the base, the triangles adjoining the perpendicular are similar both to the whole and to one another." I concurred with his interpretation and asked to see more ancient mathematical examples. In the weeks that preceded my departure to study at the American University of Beirut, my father explained to me other ancient mathematical tablets that use the method of **completing the square** to solve quadratic equations.

Among the mathematical cuneiform tablets that Taha Baqir had discovered and published are the following:

Db_2-146 published in Baqir (1962), IM 52301 in Baqir (1950b), IM 53953 in Baqir (1951), IM 53957 in Baqir (1951), IM 53961 in Baqir (1951), IM 53965 in Baqir (1951), IM 54010 in Baqir (1951), IM 54011 in Baqir (1951), IM 54464 in Baqir (1951), IM 54478 in Baqir (1951), IM 54538 in Baqir (1951), IM 54559 in Baqir (1951), and IM 55357 in Baqir (1950). Goncalves (2015) published an entire book discussing Baqir's discoveries at Tell Harmal.

Baqir studied at the American University of Beirut and the Oriental Institute of the University of Chicago. He was born in Babylon 1912, and he died in Baghdad, Iraq, in 1984.

In writing this book, I have relied mainly on several of Taha Baqir's notes and writings (written in Arabic); also, I surveyed several recent specialized publications on the subject. To jog the reader's memory, I have included reviews of some essential mathematical topics. I believe the book will be of value to both the specialists and the public.

About the Author

Saad Taha Bakir holds a Ph.D. in statistics from Virginia Tech, USA; he has MS and BS degrees in mathematical statistics from the American University of Beirut, Lebanon. Dr. Bakir worked at several universities, including Alabama State University, Virginia Tech, Yarmouk University, Kuwait University, Virginia Commonwealth University, and the University of Baghdad. He coauthored the Kuwait Arabic-English *Encyclopedia of Mathematics* and has published in several scholarly journals such as *Technometrics, Sequential Analysis, Quality Technology, Communications in Statistics*, and others. The author has a deep interest in the history of mathematics.

Saad Taha Bakir

stbakir@gmail.com

Abbreviations

AO: Antiquités Orientales (Louvre, Paris)
BM: British Museum
IM: Iraqi Museum
MCT: Mathematical Cuneiform Texts
MLC: Morgan Library Collection (deposited at Yale University, New Haven)
MS: Manuscripts in the Schoyen Collection, Oslo and London
VAT: *Vorderasiatische Abteilung, Tontafeln, Staatliche Museen*, Berlin
YBC: Yale Babylonian Collection

List of Mesopotamian tablets

	Tablet Identification	Topic
1	**AO 6484**	Progressions
2	**AO 6770**	Compound interest
3	**AO 8862**	System of equations
4	**BM 13901**	Expansion rules
5	**BM 34568**	Sliding poles on a wall
6	**BM 85194**	Pythagorean theorem
7	**BM 85196**	Progressions; sliding poles
8	**BM 85200 + VAT 6599**	Sums of cubes and squares
9	**BM 106425**	Reciprocals
10	**IM 67118 (Db_2-146)**	Simultaneous equations
11	**IM 52304**	Reduction to unity principle
12	**IM 52685**	Lists of equations
13	**IM 52916**	Lists of equations
14	**IM 55357**	Similar triangles/Pythagoras
15	**IM 58045**	Trapezoid partitioning
16	**MLC 2078**	Logarithms
17	**MS 3899**	Sums of cubes and squares
18	**Plimpton 322 (Columbia Univ.)**	Pythagorean triples
19	**YBC 4675**	Trapezoid
20	**YBC 6967**	Quadratic equations
21	**YBC 7289**	Square root of 2
22	**YBC 8512**	Triangles and trapezoids
23	**VAT 6958**	Pythagorean theorem
24	**VAT 7848**	Trapezoids
25	**VAT 8492**	Sums of cubes and squares
26	**VAT 8528**	Compound interest

Introduction

This book presents a brief history of ancient mathematics that had developed in Mesopotamia (ancient Iraq), from the early Sumerian civilization (ca. 3500 BCE) up to the fall of the Seleucid rule (125 BCE). Mesopotamia is a Greek word meaning "between two rivers." Mesopotamian mathematics has reached us as writings on clay tablets (sun-dried or baked) in the Sumerian and Akkadian languages using the cuneiform script. Because most of the excavated mathematical tablets date back to the Babylonian period, the term "Babylonian mathematics" has been used instead of "Mesopotamian mathematics." Several historians believe that many principles of present-day mathematics had originated in Mesopotamia some five thousand years ago. The Mesopotamian mathematical accomplishments can be summed up by the following quote from Eleanor Robson (2008, 1):

> Iraq-Sumer-Babylonia-Mesopotamia: under any or all of these names almost every general textbook on the history of mathematics assigns the origins of "pure" mathematics to the distant past of the land between the Tigris and Euphrates rivers. Here, over five thousand years ago, the first systematic accounting techniques were developed, using clay counters to represent fixed quantities of traded and stored goods in the world's earliest cities. Here too, in the early second millennium BCE, the world's first positional system of the numerical notation—the famous sexagesimal place value system was widely used. The earliest evidence for "pure" mathematics includes a very accurate approximation to the square root of 2, the first form of abstract algebra, and the knowledge, if not proof, of 'Pythagoras' theorem defining the relationship between the

> sides of a right-angled triangle. The best-known mathematical artifact from, this time, the cuneiform tablet Plimpton 322, has been widely discussed and admired for its function that range from number theory to trigonometry to astronomy.

In writing this book, I have often taken the time to refresh the reader's memory on a certain mathematical topic before presenting the ancient text. I hope that this book will be of value to both the public and the specialists. The book is based on several of the scattered notes and publications (mostly written in Arabic) of the renowned Iraqi archeologist Taha Baqir. Other sources include Friberg (2007), Goncalves (2015), Hoyrup (2002), Neugebauer (1945, 1969), and Robson (2008).

This book is composed of nine chapters. Chapter 1 presents a historical overview of Mesopotamia. Chapter 2 discusses the place-value number systems. Chapter 3 explains the Mesopotamian number system and compares it to our present-day decimal system. Chapter 4 discusses the discovery and types of the ancient mathematical texts. Chapter 5 presents some table texts. Chapter 6 presents a review of quadratic equations. Chapter 7 presents ancient problem texts that deal with equations. Chapter 8 presents mathematical problem texts that deal with geometric figures. Chapter 9 discusses ancient tablets that deal with the mathematics of compound interest.

CHAPTER 1 A Historical Overview of Mesopotamia

As I have mentioned in the introduction, this book presents a brief history of ancient mathematics that had developed in Mesopotamia (ancient Iraq), from the early Sumerian civilization (ca. 3500 BCE) up to the fall of the Seleucid rule (125 BCE). Iraq was known in antiquity as "Mesopotamia," which is a Greek word meaning "between two rivers." The two rivers are the Tigris and the Euphrates; they have their sources in Turkey and flow in a south-eastern direction through Mesopotamia until they merge into one river, Shat al-Arab, in southern Iraq.

The Euphrates (1727 miles long) passes close to several famous archeological sites, including (from north to south) Terqa, Mari, Sippar, Babylon, Kish, Nippur, Shuruppak, Uruk, Larsa, Ur, and Eridu. The Tigris (1211 miles long) passes (north to south) close to the archeological sites of Nineveh, Assur, Eshnuna, Tell Harmal, and Tell Dhibai. Contemporary Iraq is bordered by Iran in the east, Kuwait and Saudi Arabia in the south, Jordan and Syria in the west and Turkey in the north. Historians divide Mesopotamia into four archeological regions: **Assyria** in the north, **Akkad and Babylonia** in the middle, and **Sumer** in the south. See Fig. 1.

Archeologists divide the history of humanity into three parts: **Prehistory, Protohistory,** and **History**. **Prehistory** extends from the first appearance of human-like creatures, one million years ago, to just before the invention of primitive writing in 3700 BCE. **Protohistory** spans the time from the invention of primitive pictogram writing to the first appearance of mature writing in 2800 BCE. **History** begins with the use of mature writing for record keeping; it continues to our present day. A chronology of the main periods and events throughout the history of Mesopotamia follows.

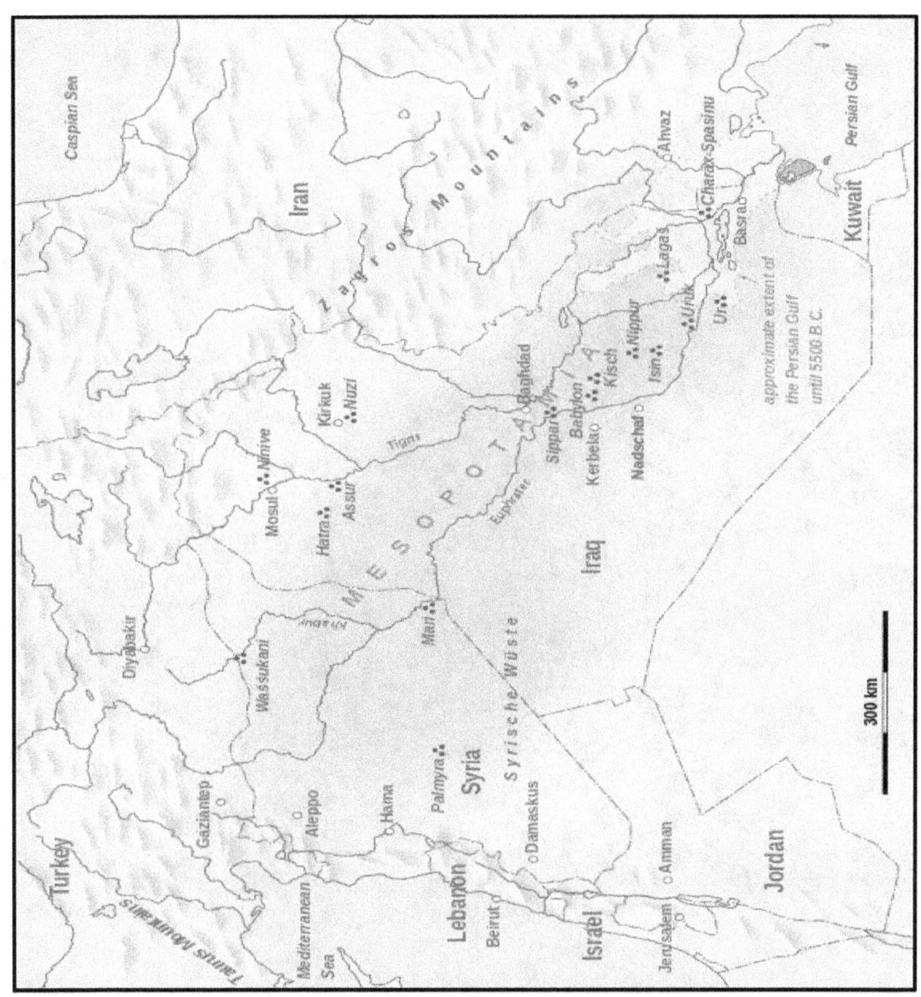

Figure 1. Map of Mesopotamia and surrounding areas; courtesy of Goran tek-en—own work based on Karte von Mesopotamien Mesopotamia Syria, CC BY-SA 3.0, https://commons.wikimedia.org/w/index.php?curid=30851043

Prehistory

In prehistoric times, Mesopotamia was home to Neanderthal cultures, such as those found in Shanidar cave in northern Iraq. Prehistory spanned about 99 percent of the life of humankind (one million to 3750 BCE). Archeologists divide prehistory into **Stone Ages** and **Stone-Metal Ages**.

Stone Ages

Paleolithic Periods: ca. 70,000–9000 BCE

The Paleolithic period spanned about 98 percent of the human history, and it was contemporary with the Ice Ages during the Pleistocene geological age. The first part of the Paleolithic was dominated by ancient species of human-like creatures; the Middle Paleolithic (ca. 60000–35000 BCE) was dominated by Neanderthal hunters living in caves. Archeological sites are Shanidar cave near present-day Rowanduz and Barda Balka near present-day Chemchemal in northern Iraq. Shanidar is a vast cave that preserved several skeletons belonging to the Neanderthal species. The Upper Paleolithic Period (ca. 35000–9000 BCE) witnessed the appearance of the ancestors of present-day humans, the *Homo sapiens*; the main archeological site of this period is Shanidar cave.

Mesolithic Period: ca. 9000–7000 BCE

This period witnessed the beginning of animal domestication, developments of small bone tools, weapons, and clay works. Archeological sites are Shanidar cave, Karim-Shehir near present-day Chemchemal in northern Iraq, and Bus-Mordeh in southern Iraq.

Neolithic Period: ca. 7000–6000 BCE

This period witnessed a monumental turning point in civilization when humans progressed in domesticating animals such as sheep and dogs, cultivating food plants such as lentils and peas, and making baked pottery and bricks. Archeological sites are Jarmo and Tell Simshara near Chemchemal and Tepe Ali Kosh in southwestern Iran.

Stone-Metal (Chalcolithic) Age: 6000–3750 BCE

During this age, the Mesopotamians employed metals in their tools and developed irrigation agriculture. The periods of this age are named after their geographic sites as follows:

The Hassuna Period (5800–5500 BCE): The archeological site is Tell Hassuna near present-day Mosul in northern Iraq.

The Samarra Period (5600–5000 BCE): The archeological site is Samarra in central Iraq.

The Halaf Period (5500–4500 BCE): The archeological site is on the Khabur River near the Iraq-Turkey border.

The Ubaid 1 and 2 Periods (5000–3750 BCE): The archeological site is near the Sumerian city of Ur (near present-day Nassyria in southern Iraq). This period witnessed increased urbanization including the erection of temples and houses of large sizes; villages had expanded into cities and settlements flourished in the alluvial valleys of middle and southern Iraq.

Protohistory: 3750–2800 BCE

This age marked the beginning of civilization in the world with the invention of primitive writing in the form of pictograms. However, writing was not yet used for record keeping purposes. Periods of this age are:

The Uruk Period: 3750–3150 BCE

The archeological site of this period is Uruk (Warka or Erech) near present-day Samawa in southern Iraq. This period marked the real birth of civilization with the use of writing for record keeping; urbanization and trade had expanded. Several sculptures of bronze, gold, and silver have been excavated at the Uruk site.

The Jemdat Nasr Period: 3150–2900 BCE

The archeological site of this period lies between Baghdad and Babylon in central Iraq. This period witnessed the beginning of city-state governments, construction of *ziggurats*, and improvement of irrigation systems. Moreover, painted pottery and sculptures have been excavated from this period.

History: 2800 BCE - Present

This age marks the time when cuneiform writing matured and was used for record keeping of life affairs. Local government structures evolved into dynastic rules at various city-states. We present now a brief political history of Mesopotamia.

Early Dynastic (City-States) Periods: 2800-2370 BCE

The Early Dynastic Periods were characterized by city-state type of governments where a ruler resides in a large city and governs several satellite towns and villages. During this period, cuneiform writing matured and was used for record keeping of miscellaneous life affairs and governmental laws. Several competing and ethnically distinct dynasties—the non-Semite Sumerians and the Semite Akkadians—ruled at various locations in southern Mesopotamia. Early Dynastic I and II Periods were established at Kish (near present-day Hilla in southern Iraq) and Uruk (near present-day Samawa in southern Iraq.) Early Dynastic III Period witnessed some strong Sumerian and Akkadian dynasties at Kish, Uruk, Ur, and Lagash (near present-day Nassyria in southern Iraq).

Various ethnic groups dominated the political history of Mesopotamia. Those groups, in chronological order, were the Sumerians, Akkadians, Amorites (Old Babylonians), Hittites, Kassites, Assyrians, Aramaeans, Chaldaeans (Neo-Babylonians), Achaemenids Persians, Seleucids (Hellenistic), Parthian Persians, Sassanid Persians, and the Arabs/ Muslims. For a detailed account of the history of Mesopotamia, see Baqir (1973, written in Arabic) and Roux (1992).

The Sumerians (2800–2370 BCE)

The Sumerians were non-Semitic people of unknown origin; they ruled throughout the Early Dynastic Periods (2800–2370 BCE). Several competing dynasties, the Semite Akkadians, ruled concurrently at other city-state locations in the lower part of Mesopotamia. Early Dynastic I and II Periods were established at Kish and Uruk. One famous Sumerian king at Uruk was Gilgamesh, the hero of the famous epic of Gilgamesh. Early Dynastic III Period witnessed several strong Sumerian and Akkadian dynasties at Kish, Uruk, Ur, and Lagash. After the time of King Lugalzagesi (2340–2316 BCE), the Sumerian land fell at the hands of King Sargon I who established a vast Akkadian empire.

The Akkadians: 2370–2160 BCE

The Akkadians were Semitic people who migrated northward from the Arabian Peninsula and established city-states alongside the Sumerian city-states in southern Iraq. The famous king of Akkad (or Agade), Sargon I then conquered all the Sumerian city-states and established an Akkadian Empire (2370–2160 BCE) over most of Mesopotamia. During the rule of King Naram-Sin, grandson of Sargon I, the Akkadian Empire expanded to rule over several parts of the Near East; it lasted for a century. After the death of Naram-Sin, the Akkadian Empire fell under the hands of the Gutians who came from the Zagros Mountains in Iran.

The Gutians: 2230–2120 BCE

Migrating from the Zagros Mountains in Iran, the Gutians invaded Sumer and Akkad; thus they destroyed the Akkadian Empire. Not much is known about the Gutians rule, except it lasted for a hundred years and was plagued by anarchy and chaos.

The Sumerians Rise Again (2140–2110 BCE)

A Sumerian dynasty rose and established an independent rule at Lagash, contemporary with the Gutians period. Several writings and statutes of its famous king, Gudea (2140–2120 BCE), have been excavated. The last of the Sumerians, Ur Dynasty III, ruled at Ur for the period 2112–2004 BCE. This dynasty established a strong kingdom that governed vast regions of Mesopotamia and the Near East; most famous kings of this dynasty were Ur-Nammu (2112–2095 BCE) and Ibb-Sin (2028–2004 BCE.) The Sumerian Era ceased to exist when the city of Ur fell around 2004 BCE to the Amorites. Nevertheless, the Sumerian language continued to be in use for writing and record keeping all over Mesopotamia.

The Amorites at Isin and Larsa: 2000–1595 BCE

The Amorites were Semitic nomadic tribes who lived in the Syrian valleys west of Mesopotamia up to the Mediterranean Sea. Vast migrations of the Amorites had established dynasties at Isin and Larsa in southern Mesopotamia.

Dynasty at Isin: 2017–1794 BCE

Some Amorite tribes established a dynastic rule at Isin, an ancient city about twenty miles south of Nippur (the present-day Niffar or Nuffar) in southern Iraq. One king of this period, Libit-Ishtar (1934–1924 BCE), became famous for writing a code of law that deals with succession, real estates, and work contracts. In 1787 BCE, the Isin dynasty fell at the hands of the Babylonian king, Hammurabi.

Dynasty at Larsa: 2025–1763 BCE

Other Amorite tribes established a dynastic rule at Larsa, an ancient city about fifteen miles southeast of Uruk in southern Iraq. Sixteen Amorite kings ruled during this period. In 1763 BCE, Larsa fell at the hands of the Babylonian king, Hammurabi.

Dynasties at Babylon (the Babylonians)

In total, about eleven dynasties ruled over the city of Babylon (near present-day Hillah in southern Iraq). Babylonians spoke the Akkadian language and retained the Sumerian language for religious affairs; their culture was a synthesis of Akkadian and Sumerian cultures.

Babylon Dynasty I (the Old Babylonian Period): 1894–1595 BCE

A strong Amorite tribe settled in Babylon and established a city-state rule. Its sixth king, Hammurabi (1792–1750 BCE), conquered all the city-states of Sumer and Akkad and established a vast Babylonian Empire encompassing all Mesopotamia and many Near Eastern regions. King Hammurabi is famous for his "code of law" that consists of 282 sections dealing with most aspects of business and social life. The laws are inscribed on a hard stone slab that currently resides in the Louvre Museum in Paris, France. This period (called the Old Babylonian period) is distinguished by the emergence of vast activities of writing and record keeping. Sumerian literature and laws were translated into the Semitic Babylonian language. Most importantly, the Old Babylonian period has provided us for the first time with many ancient mathematical texts, including table text (multiplication, square roots, and exponents) and problem texts dealing with algebraic and geometric topics. In 1595 BCE, the Babylonian rule fell at the hands of the Hittites (Peoples from Anatolia).

Babylon Dynasty II (Kings or Peoples of the Sea): 1900–1525 BCE

King Iluma-ilu, a descendant of Damiq-ilishu (the last king of Isin), established Babylon Dynasty II. Twelve kings of this dynasty had ruled over the Gulf region south of Nippur. Most probably, Peoples of the Sea originated from the Aegean isles.

The Hittites

The Hittites were Indo-European Peoples from Asia Minor (Anatolia). They invaded Mesopotamia and sacked both Akkad and Babylon. For unknown reasons, however, the Hittites soon left Babylon and returned home.

Babylon Dynasty III (the Kassites): 1500–1156 BCE

The Kassites—Indo-Europeans from Iran—established a dynasty rule (Babylon Dynasty III) that lasted for about four centuries. Among other things, they introduced the horse to Mesopotamia. During this period, translation of literal and scientific works flourished, and Babylonia became a prosperous region again. The Kassites' rule fell at the hands of the Elamites—people from southern Iran.

The Assyrians: 2000–605 BCE

The Assyrians were Semitic Peoples who established their strong regional kingdoms at Assur, an ancient site near the present-day Qala't Sherqat in northern Iraq. Assur had always been a trading post between the southern and northern regions of Mesopotamia. The Assyrian land had been ruled by several independent dynasties until it was unified into one strong empire in the tenth century BCE, and Babylon became part of the Assyrian Empire in the eighth century BCE. Among the main Assyrian cities were **Assur**, **Kalhu (Nimrud)**, and **Nineveh**. Historians divide the Assyrian rule into **Old-**, **Middle-**, and **Neo-Assyrian** Periods.

The Neo-Assyrian Empire

This empire began with the rule of King Adad-nirari II in 911 BCE, and it was vastly expanded by the famous Assyrian king, Sargon II (722–705 BCE). His son, Sennacherib (705–681 BCE), defeated the Greeks, drove the Egyptians out of Israel and Phoenicia, and crushed a Babylonian revolt. King Esarhaddon (680–669 BCE), son of Sennacherib, expanded Assyria further into the Caucasus Mountains and fought the Egyptians in the Sinai Desert. The empire reached as far south as Arabia and Dilmun (present-day Bahrain and Qatar). Esarhaddon rebuilt Babylon and brought peace to all Mesopotamia.

Under Ashurbanipal (669–627 BCE), son of Esarhaddon, the Assyrian empire spanned from the Caucasus Mountains in the north to Egypt and Arabia in the south and extended from Cyprus in the west to Persia in the east. After the death of Ashurbanipal, Assyria descended into a series of civil wars, and Nineveh fell at the hands of the Neo-Babylonians (the Chaldeans) in 605 BCE.

The Neo-Babylonian Empire (Chaldaeans): 626–539 BCE

In 626 BCE, Nabopolassar (of a dominant Chaldaean tribe), established a new dynastic rule in Babylon. Nabopolassar's son, Nebuchadnezzar II (605–562 BCE), expanded his rule to a large empire that included Jerusalem (586 BCE), from which he seized thousands of captives into exile in Babylon. King Nebuchadnezzar II is most famous for building the Hanging Gardens of Babylon. The last Babylonian kings were Nabonidus and Belshazzar (550–539 BCE). In 539 BCE, Babylonia fell at the hands of the Achaemenid Persians who were led by King Cyrus II.

The Achaemenid Persians: 539–331 BCE

Named after their leader, Achaemenid, the Achaemenids established their rule in central Iran around 700 BCE. They expanded to a vast empire under King Cyrus II (or Kourosh, 559–529 BCE), who conquered Babylon in 539 BCE and invaded the Syrian region thereafter. Cyrus II is famous for writing the first human rights declaration as inscribed in the Cyrus Cylinder. Another king, Darius I (521–486 BCE), ruled over a vast Persian empire stretching from Afghanistan in the east to Egypt and some Greek cities in the west.

King Darius I left a paramount monument (records of events and victories of his rule) that helped archaeologists decipher the cuneiform writing. Henry Rawlinson (1810–1895 CE), a British officer, discovered a rock at the Zagros Mountains that is inscribed in three ancient languages: Akkadian/Babylonian, Elamites (the language of the Achaemenids Empire), and Old Persian. This site is referred to as the "Behistun Inscriptions" because it is located in the town of Behistun (or Bisotun) along the road from Hamadan in Iran to Babylon in Iraq. Rawlinson managed to decipher the Old Persian inscriptions first; deciphering the cuneiform writing then followed. During the rule of King Darius III (336–330 BCE), the Achaemenids Empire fell in 330 BCE at the hands of Alexander the Great of Macedonia.

The Hellenistic Period: Alexander the Great and the Seleucid Dynasty (323–126 BCE)

After defeating the Achaemenids in 330 BCE, Alexander the Great entered Babylon and advanced further east to reach the Indus River. After nine years of conquest, Alexander turned back and died, out of sickness, in Babylon in 323 BCE at the age of thirty-two. Subsequently, the empire was divided among his generals, and Mesopotamia was allotted to general Seleucus and his descendants. Some ancient mathematical tablets came to us from the Seleucid period. The Seleucids rule ceased when the Parthian Persians conquered Babylon in 126 BCE and dominated Mesopotamia.

The Parthian Persians: 140 BCE–227 CE

The Parthians (or the Arsacids, after their leader Arsaces), were nomadic tribesmen from northeastern Iran. In 140 BCE, they conquered all of Iran and established themselves at the city of Ctesiphon (on the present Iraqi-Iranian border). In 126 BCE, their king, Artabanus, conquered Babylon and seized all Mesopotamia of the Seleucids' domain. The Parthians rule fell at the hands of the Sassanids in 227 CE. Incidentally, there were two brief periods of Roman rule over Mesopotamia by the Roman emperors Trajan and Severus.

The Sassanid Persians: 226–637 CE

The Sassanids, named after their high priest Sassan, originally lived in southern Iran. Under their king, Ardashir I (224–40 CE), they defeated the Parthians, took over Ctesiphon in 226 CE, and occupied most of Mesopotamia. King Shapur I (240–72 CE) conquered some Roman territory, and he imprisoned the Roman emperor, Valerian, who later died in Persian captivity. The Sassanids kept their empire strong until 637 CE when they were defeated by the Muslim Arabs at the battle of al-Qadisiyah. Ctesiphon and all the Sassanid's territories were later seized by the Muslims in 638 CE during the rule of the Caliph Umar ibn-al-Khattab (634–44 CE). The last Sassanid king, Yazdegerd III, fled to northern Persia and was later killed in 651 CE.

The Muslim Caliphates and Dynasties

After the defeat of the Sassanids, Mesopotamia was ruled by the following Muslim caliphates or dynasties:

The Patriarchal (or Orthodox) Caliphs (Abu-Bakr, Umar, Othman, and Ali): 632–61 CE.

The Umayyad Caliphate (661–750 CE)

The Abbasid Caliphate (750–1258 CE)

The Mongol invasion and the fall of Baghdad, 1258 CE

Some Mongol and Turkoman Dynasties (1258–1508 CE)

The Persian Safavid Dynasty (1508–1523 CE) and (1529–34 CE)

The Ottoman Turks Empire (1534–1917 CE).

After the collapse of the Ottoman Empire in World War I, Iraq became a kingdom for the period 1921–58 CE. After that, Iraq has been a republic.

CHAPTER 2 Number Systems

To understand the mathematics of a particular civilization, we must first discern the nature of the number system of that civilization. The Mesopotamians used the base-60 (sexagesimal) system for most scientific purposes and sometimes used the base-10 (decimal) system for administrative purposes. Other civilizations, such as the Mayan, used a base-20 system. Our present-day culture uses the base-10 system for most calculations in addition to using the base-60 system for time and trigonometric measurements. Present-day computer scientists use bases 2, 8, and 16 in their applications.

To help understand the Mesopotamian mathematical texts, we discuss in this chapter the sexagesimal number system and its relation to our present-day decimal number system.

Place-Value Number Systems

In general, a place-value (or positional-value) number system has the following components:

1. A base (or radix) *b*, which is an integer (whole number) greater than 1.
2. Characters to represent the numerals between 1 and (b – 1) inclusive.
3. A character to indicate an empty place value (the zero in our present-day system).
4. A character (a separator) to separate the integral and the fractional parts of a number.

Let b be the base, the p_i's be the numerals including an empty place character, and $\$$ be the fractional separator of a number system. Consider a number Z written as $p_k p_{k-1} p_{k-2} \cdots p_1 p_0 \$ p_{-1} p_{-2} p_{-3} \cdots$.

Then, in this base-b system, Z has the value:

$$Z = p_k \times b^k + p_{k-1} \times b^{k-1} + p_{k-2} \times b^{k-2} + \cdots p_1 \times b^1 + p_0 \times b^0$$

$$+ p_{-1} \times b^{-1} + p_{-2} \times b^{-2} + p_{-3} \times b^{-3} + \cdots$$

$$\text{Note}: b^0 = 1 \text{ and } b^{-m} = \frac{1}{b^m}.$$

The Decimal and the Sexagesimal Place-Value Number Systems

The present-day decimal place-value number system uses base $b = 10$ and it employs the Arabic numerals 1, 2, 3, 4, 5, 6, 7, 8, and 9 to represent the numbers between 1 and 9 (= b − 1). It uses "0" for an empty place and the decimal point "." for separating the integral and the fractional parts of a number.

The sexagesimal place-value number system uses base $b = 60$ and it employs the Arabic numerals 1, 2, 3, 4, 5, 6, 7, 8, 9, 10, 11,..., 59 for the numbers between 1 and 59 (= b − 1). It uses "0" for an empty place. Also, the sexagesimal system needs a symbol for separating the place values because a location may hold a number consisting of two characters, such as 59. We will follow Neugebauer and Sachs (1945) notation to use the comma "," for separating the place-value positions and the semicolon ";" for separating the integral and the fractional parts of a sexagesimal number. Thus, the semicolon ";" is the sexagesimal point that corresponds to the dot "." in the decimal point.

Example: The number 321$04 (where $ is the fraction separator) has entirely different values in the decimal and the sexagesimal systems. To distinguish the two representations, we write the intended base as a subscript to the number. In base 10, the number 321$04 is written as 321.04_{10} and it has value:

$$\begin{aligned}321.04_{10} &= 3\times 10^{2} + 2\times 10^{1} + 1\times 10^{0} + 0\times 10^{-1} + 4\times 10^{-2} \\ &= 300 + 20 + 1 + 0 + 0.04 = 321.04\end{aligned}$$

In base 60, the number 321$04 is ambiguous, because it may mean $3,2,1;0,4_{60}$ or $32,1;0,4_{60}$, etc. These expressions the following values:

$$\begin{aligned}3,2,1;04_{60} &= 3\times 60^{2} + 2\times 60^{1} + 1\times 60^{0} + 0\times 60^{-1} + 4\times 60^{-2} \\ &= 10800 + 120 + 60 + 1 + 0 + .0011 = 10981.0011\end{aligned}$$

$$\begin{aligned}32,1;0,4_{60} &= 32\times 60^{1} + 1\times 60^{0} + 0\times 60^{-1} + 4\times 60^{-2} \\ &= 1920 + 1 + 0 + .0011 = 1921.0011\end{aligned}$$

Converting Numbers from Base 60 to Base 10

To convert a base-60 number to a base-10 number, just multiply the numerals by the relative powers of 60, and add up the results.

Examples:

$$\begin{aligned}5,11;2,24_{60} &= 5\times 60^{1} + 11\times 60^{0} + 2\times 60^{-1} + 24\times 60^{-2} = \\ &= 300 + 11\times 1 + 2\times\frac{1}{60} + 24\times\frac{1}{3600} = 311.04_{10}\end{aligned}$$

$$1,24,51,10_{60} = 1\times 60^{3} + 24\times 60^{2} + 51\times 60 + 10\times 60^{0}$$

$$= 216000 + 86400 + 3060 + 10 = 305470_{10}$$

The above technique works for converting any base-*b* number to a base-10 number: just multiply by the powers of the intended base instead of 60.

Examples:

$$14236_8 = 1\times 8^4 + 4\times 8^3 + 2\times 8^2 + 3\times 8^1 + 6\times 8^0$$

$$= 1\times 4096 + 4\times 512 + 2\times 64 + 3\times 8 + 6\times 1$$

$$= 4096 + 2048 + 128 + 24 + 6 = 6302_{10}$$

$$14{,}2{,}3{,}6_{16} = 14\times 16^3 + 2\times 16^2 + 3\times 16^1 + 6\times 16^0$$

$$= 14\times 4096 + 2\times 256 + 3\times 16 + 6\times 1$$

$$= 57344 + 512 + 48 + 6 = 57910_{10}$$

Note that present-day computer scientists use bases 2, 8, and 16 in some applications.

The following are useful online number system converters

---

http://baptiste.meles.free.fr/site/mesocalc.html

http://convertxy.com/index.php/numberbases/

http://www.easysurf.cc/cnver17.htm (works for any base and accepts fractions)

CHAPTER 3 Mesopotamian Number System

In this chapter, we discuss the Mesopotamian number system in order to understand the mathematics of the ancient tablets.

The Mesopotamian number system has the following characteristics:

1. It is a place-value system with base $b = 60$.
2. It uses only two wedge-type characters for writing the numerals: "Y" for representing 1 or 60, and "<" for representing 10.
3. It has no particular character to indicate an empty place value (zero). Usually, a gap left between the numerals indicates a zero. Only later, in the Seleucid period (third century BCE), a character was adopted to represent zero.
4. It has no character (a separator) to separate the integral and fractional parts of a number. Usually, a gap left between the numerals indicates a fractional separator.

The Mesopotamians wrote their numbers by a repetitive inscription of the two characters "Y" for 1 or 60 and "<" for 10 side by side or on top of each other. The absence of a symbol for zero and for a separator led to some uncertainty in figuring out the intended values of some cuneiform numbers. Experts often had to rely on the contents of the mathematical cuneiform texts to resolve those ambiguities. Because the base 60 is divisible by more factors than the base 10, the sexagesimal number system provides more flexibility, especially, in writing fractions. The fifth (1/5), the quarter (1/4), the third (1/3), and the two-thirds (2/3) of our present-day decimal system are written, respectively, as 0;12 0;15 0;20, and 0;40 in the sexagesimal number system. We will use a comma "," to separate positions of the powers of 60 and use a semicolon ";" to separate the integral and the fractional parts of a sexagesimal number. We will keep using "0" for empty positions. As examples, we take the numbers discussed in the previous chapter and write them in cuneiform characters.

Example: The cuneiform representation of $311.04_{10} = 5,11;2,24_{60}$ is

$\text{cuneiform}\begin{matrix}YYY\\YY\end{matrix} < Y \quad YY << \begin{matrix}YYY\\Y\end{matrix}$, which is inscribed on tablet IM 55357 that deals with similar triangles and the Pythagorean Theorem.

Note the gap in the middle of the cuneiform number; it indicates the beginning of the fractional part of the number.

Example: The cuneiform representation of $424000_{10} = 1,57,46,40_{60}$ is

$$Y \quad \begin{matrix}<<<\\<<\end{matrix}\begin{matrix}YYY\\YYY\\Y\end{matrix} \quad \begin{matrix}<<<\\<\end{matrix}\begin{matrix}YYY\\YYY\end{matrix} \quad \begin{matrix}<<<\\<\end{matrix}$$

Tables 1 and 2 display the cuneiform representations of some whole and fractional numbers.

Table 1. Number Representations

Decimal	Cuneiform	Sexagesimal
0	No symbol	0
1	Y	1
2	YY	2
3	YYY	3
4	YYY Y	4
5	YYY YY	5
9	YYY YYY YYY	9
10	<	10
11	<Y	11
20	<<	20
21	<<Y	21
60	Y	1,0
61	YY	1,1
70	Y<	1,10
100	Y << <<	1,40
120	YY	2,0
151	YY<<<Y	2,31
400	YYY << YYY <<	6,40
44733	< YY < YYY <<< YYY YY	12,25,23

Table 2. Fractional Representations

Common Fraction	Sexagesimal	Value
0	0	0×60^0
1/3600	0;0,1	$0\times 60^0 + 0\times 60^{-1} + 1\times 60^{-2}$
1/120	0;0,30	$0\times 60^0 + 0\times 60^{-1} + 30\times 60^{-2}$
1/60	0;1	$0\times 60^0 + 1\times 60^{-1}$
1/50	0;1,12	$0\times 60^0 + 1\times 60^{-1} + 12\times 60^{-2}$
1/40	0;1,30	$0\times 60^0 + 1\times 60^{-1} + 30\times 60^{-2}$
1/30	0;2	$0\times 60^0 + 2\times 60^{-1}$
1/25	0;2,24	$0\times 60^0 + 2\times 60^{-1} + 24\times 60^{-2}$
1/20	0;3	$0\times 60^0 + 3\times 60^{-1}$
1/16	0;3,45	$0\times 60^0 + 3\times 60^{-1} + 45\times 60^{-2}$
1/15	0;4	$0\times 60^0 + 4\times 60^{-1}$
1/12	0;5	$0\times 60^0 + 5\times 60^{-1}$
1/10	0;6	$0\times 60^0 + 6\times 60^{-1}$
1/9	0;6,40	$0\times 60^0 + 40\times 60^{-1}$
1/8	0;7,30	$0\times 60^0 + 7\times 60^{-1} + 30\times 60^{-2}$
1/6	0;10	$0\times 60^0 + 10\times 60^{-1}$
1/5	0;12	$0\times 60^0 + 12\times 60^{-1}$
¼	0;15	$0\times 60^0 + 15\times 60^{-1}$
1/3	0;20	$0\times 60^0 + 20\times 60^{-1}$
½	0;30	$0\times 60^0 + 30\times 60^{-1}$
$1\frac{1}{2}$	1;30	$1\times 60^0 + 30\times 60^{-1}$
$6\frac{2}{3}$	6;40	$6\times 60^0 + 40\times 60^{-1}$

CHAPTER 4 Mesopotamian Mathematics

In this chapter, we give an overview of the discovery, types, and characteristics of the Mesopotamian mathematical tablets.

Discovery

The decipherment of the cuneiform writing began during the mid-nineteenth century CE with the work of Henry Rawlinson on the Behistun Inscriptions. Not until the late 1920s, however, that cuneiform experts realized the depth of the mathematical contents of so many of the excavated cuneiform tablets. They perceived texts with numerals as mere records of revenues or simple arithmetical tables. The situation changed when Otto Neugebauer (1929) published the first mathematical cuneiform tablet. Most of Neugebauer's work appeared in *Mathematische Keilschrifttexte* (1935–38), and in his 1945 monumental book (with A. Sachs and A Goetze), *Mathematical Cuneiform Texts*. The second pivotal researcher was Thureau-Dangin, who published most of his work in *Mathematiques Babyloniens* (1938). Then, E. M. Bruins published most of his work in the Iraqi journal, *Sumer* (1950–55). During 1945–62, Taha Baqir (the renowned Iraqi archeologist) excavated hundreds of cuneiform tablets in Tell Harmal and Tell Dhibai, two ancient sites near present-day Baghdad. In the early 1950s, Baqir published in Sumer twelve of the Tell Harmal mathematical tablets. The Tell Harmal discovery has continued to attract scholars' attention to recent times. In 2015, Carlos Goncalves published an entire book on those twelve tablets that Baqir had discovered. Other recent books on mathematics in Mesopotamia include Friberg (2007), Hoyrup (2002), and Robson (2008).

The scholars mentioned above believe that several principles of our present-day mathematics date back to Mesopotamia, some five thousand years ago. There is evidence that the Mesopotamians had used some fundamental theorems and laws in geometry and algebra some seventeen hundred years before the famous Greek mathematicians such as Euclid and Pythagoras.

Most of the excavated mathematical tablets date back to one of two historical periods of Mesopotamia:

1. The Old Babylonian period (2000–1500 BCE)
2. The Seleucid Period (323 - 126 BCE).

None of the excavated tablets, except for a few containing numerical symbols, dated back to periods before the Old Babylonian. Did mathematics suddenly flourish around 2000 BCE in Mesopotamia? The answer is still a historical puzzle with no satisfactory answer.

Types of Mesopotamian Mathematics

Historians classify the mathematical texts tablets that reached us from Mesopotamia into two main groups: "problem texts" and "table texts." The problem texts group contains a variety of algebraic and geometrical problems involving quadratic equations, progressions, and some geometric figures. The table texts group includes arithmetical tables of multiplications, squares, roots, reciprocals, and some special coefficients. In addition to these two groups, other mathematical texts include meteorological and astronomical tables.

Group 1: Table Texts

Table texts demonstrate a fondness for numbers and number properties. These tables include multiplications, reciprocals, roots, and some particular coefficient tables. Some strange tables have Pythagorean and logarithmic numbers inscribed on them.

Group 2: Problem Texts

Problem texts have mathematical problems written in a narrative presentation that gives the data, the question, and a step-by-step worked-out solution. Archeologists excavated about two hundred intact problem texts tablets, most of which were written in Akkadian/Babylonian, with a few in Sumerian. Several of the discovered problem texts demonstrate that Mesopotamians handled many geometrical problems in addition to linear, quadratic, and cubic equations. We can safely assert that the origin of several ideas in algebra and geometry date back to the ancient civilization of Mesopotamia. Other mathematical texts tablets contained listings of equations organized from simple standard quadratic types to more complicated non-standard types.

Characteristics of Mesopotamian Mathematics

Before discussing the actual contents of some mathematical text tablets, we describe some key characteristics of the Mesopotamian mathematics.

Fondness for Algebra

Greek mathematicians had a fondness for geometry, rather than algebra, in solving mathematical problems. In contrast, Mesopotamian mathematicians had a fondness for algebra, rather than geometry. Five thousand years ago, the Mesopotamians developed principles that laid the foundations of algebra as we know it today. Their approach of blending the figure (geometry) with the number (algebra) is what characterizes the inception (seventeenth century CE) of analytic geometry by Descartes and Fermat. In contrast, Greek mathematicians concentrated almost entirely on the study of the figure (geometry) at the expense of the number (algebra); they preferred to solve algebraic problems through geometrical arguments. An outstanding example that contrasts the algebraic path of Mesopotamian mathematics with the geometric path of Greek mathematics is the interpretation of the Pythagorean rule that relates the squares of the sides of a right triangle. Greek mathematicians treated Pythagoras' rule as a relation between geometrical figures: The square constructed on the hypotenuse of a right triangle equals the sum the squares constructed on the other two sides. In contrast, Mesopotamian mathematicians treated it as a relation between the numerical magnitudes of the sides of a right triangle. Mesopotamians even did not hesitate to add (or subtract) the side length of a square to (or from) its area, thus generating a second-degree equation as in the narrative: "The side length of a rectangle was added to its area to produce $\frac{3}{4}$ (= 0;45 sexagesimal). What is the side length?" In present-day mathematics, this narrative generates the second-degree equation $x^2 + x = \frac{3}{4}$. Geometric concepts were secondary in Mesopotamian mathematics; numbers were primary. Greek mathematicians, such as Pythagoras and his mathematical school of philosophy, focused primarily on geometric concepts and treated numbers as sophist or metaphysical entities. The Greek treatment of algebra came later during the Hellenistic era at the hands of Heron (first century CE) and Diophantus (third century CE). The work of Diophantus resembled Mesopotamian algebra to the degree that some historians attributed it to a Mesopotamian origin. Some historians opine that if the Greek mathematicians were to follow the path of the Mesopotamians, one thousand years would have been saved in the evolution of mathematics.

Non-symbolic Rhetoric Algebra

Mesopotamian mathematicians did not use alphabet letters (or symbols) to represent unknown quantities as we do in present-day mathematics. The steps of their solutions were prescriptive or rhetoric using meaningful words such as "length of a field" for the unknown and "area of a square field" to stand for the square of the unknown. Such usage resembles al-Khwarizmi's (780-850 CE) terminologies of "thing or root" for the unknown and "al-Mal" for the square of the unknown. The absence of symbols in their algebra led the Mesopotamians to invent masterful algebraic schemes to solve equations such as completing the square, substitution, elimination, false position, and the auxiliary variable schemes.

Absence of Formulae or Proofs

The Mesopotamian mathematical texts contained no proofs or formulae explaining the steps they prescribed in the solutions. The mathematical texts that have reached us thus far were written in a prescribed rhetoric format that provides the given data (information) and the question first, and then gives the steps of the worked-out solution. The consensus, however, is that those ancient mathematicians must have had at their disposal some documented mathematical laws and rules which they had discovered through investigation and proofs. Otherwise, the solutions they provided for various types of problems could not have come up by guessing or chance. Unfortunately, no ancient text that has reached us thus far contains mathematical proofs or formulae. Some historians conjectured that the ancient texts that reached us were school exercises solved by students or trainees based on theoretical principles and rules (without stating) they knew at that time. One text in the British Museum (Saggs 1962, 451, 453) may contain a theoretical proof of relations between the areas of some geometrical figures.

Main Mathematical Achievements

The mathematical tablets discovered thus far demonstrate that the Mesopotamians achieved great success in dealing with complicated mathematical topics equivalent to our present-day pre-calculus mathematics. These topics include the following:

1. Inventing the place-value number system with base 60 and extending its use to express fractions; however, a particular symbol for zero had not been in use until the Seleucid period.
2. Preparing tables for multiplication, reciprocals, exponents, square and cube roots of numbers
3. Knowledge and practical usage of the Pythagorean rule (seventeen hundred years before Pythagoras). The evidence is provided in many tablets including BM 34568, BM 85166, BM 85196, BM 85194, IM 55357, VAT 6598, Susa XIX, and Plimpton 322.
4. The solution of first-, second-, third-, and special sixth- and eighth-degree equations; they used masterful methods to solve quadratic equations, including the methods of completing the square, substitution, reduction, false position, and auxiliary variable.
5. Knowledge and use of some expansion formulae (e.g., tablet BM 13901) such as:

$$(a+b)^2 = a^2 + b^2 + 2ab, \quad (a-b)^2 = a^2 + b^2 - 2ab,$$

and $a^2 - b^2 = (a+b)(a-b)$.

They used the identity $ab = \frac{(a+b)^2 - (a-b)^2}{4}$ to aid in multiplication.

6. Working with irrational numbers: In tablet YBC 7289, the Mesopotamians calculated $\sqrt{2}$ = 1;24,51,10 (sexagesimal), which equals the decimal value of $1 + \frac{24}{60} + \frac{51}{60^2} + \frac{10}{60^3} = 1.41421296$. Also, they approximated $\sqrt{3}$ by $1\frac{3}{4}$. The irrational number π (ratio of the circumference to the diameter of a circle) was approximated by $3\frac{1}{8}$.
7. Vast geometric knowledge including the properties of similar triangles, parallel lines, areas of plane figures such as circles, parallelograms, rectangles, rhombus, trapezoids, and triangles; they worked with volumes of solid figures, such as the pyramid, frustum, prism, cylinder, cone, and truncated solid figures.
8. Calculating very mature astronomical measurements such as lengths of days and nights, according to seasons, and the lengths of the lunar and solar years.
9. Handling arithmetic and geometric progressions.
10. Calculating compound interest on loans and investments as inscribed on tablets AO 6770 and VAT 8528; see chapter 9.

CHAPTER 5 Table Texts

As we have mentioned earlier, the cuneiform mathematical tablets are classified into table texts and problem texts. In this chapter, we discuss the types and significance of the table texts tablets. Mesopotamians used these tables extensively for mathematical and astronomical calculations. The types of the table texts can be arranged into the following groups:

Tables of Multiplications

Many tables, including long multiplications of large numbers, have reached us. Some multiplication tables include a column for reciprocals too. The term for multiplication is "nasum" or "il," but no specific symbol was used. Mesopotamians used words and idioms to express addition, subtraction, multiplication, reciprocation, roots, and powers of numbers. See Neugebauer (1945, 19).

Tables of Reciprocals

Mesopotamians used to write the reciprocals in a specific field in a composite multiplication table. In our present-day decimal system, the reciprocal of a number *a* is $\frac{1}{a}$; reciprocals are pairs whose product is one. In the Mesopotamian sexagesimal system, the reciprocal of *a* is $\frac{60}{a}$; reciprocals are pairs whose product is 60. They accomplished division by multiplying by the reciprocal of the divisor: $\frac{a}{c} = a \times \frac{60}{c}$. The term for reciprocal is "igi-n-gal," which translates to "reciprocal of n."

Reciprocals of numbers may or may not be terminal expressions. For example, the reciprocal of nine in the sexagesimal system, 60/9 = 6;40, is terminal. In the decimal system, however, the reciprocal of nine, $\frac{1}{9} = 0.111\cdots$, is not terminal. Reciprocal of 7 is not terminal both in the sexagesimal the decimal system. See Neugebauer (1945, 11), Baqir (2013, 36), and Robson (2008, 81). Table 3 displays reciprocals of selective numbers inscribed on tablet BM 106425.

Table 3. Tablet BM 106425: Sexagesimal Reciprocals

a	Reciprocal $\frac{60}{a}$	a	Reciprocal $\frac{60}{a}$
2	30	25	2;24
3	20	27	2;13,20
4	15	30	2
5	12	36	1;40
6	10	40	1;30
8	7;30	45	1;20
9	6;40	50	1;12
10	6	54	1;06,40
12	5	60	1
15	4		
16	3;45		
18	3;25		
20	3		
24	2;30		

Tables of Square Roots and Cube Roots

The Mesopotamians used the idiom "IB-SI" to indicate square roots. Examples:

1	1 IB-SI	meaning $\sqrt{1} = 1$
4	2 IB-SI	meaning $\sqrt{4} = 2$
9	3 IB-SI	meaning $\sqrt{9} = 3$

For cube roots, the Mesopotamians used idiom "BA-SI." Examples:

1	1 BA-SI	meaning $\sqrt[3]{1} = 1$
8	2 BA-SI	meaning $\sqrt[3]{8} = 2$
27	3 BA-SI	meaning $\sqrt[3]{27} = 3$

Tables of Sums of Squares and Cubes

These tables give the values of $n^2 + n^3$ for various integral values of *n*. Examples:

2	1	ba.si	meaning $1^2 + 1^3 = 2$
12	2	ba.si	meaning $2^2 + 2^3 = 12$
36	3	ba.si	meaning $3^2 + 3^3 = 36$
80	4	ba.si	meaning $4^2 + 4^3 = 80$

For further reading on such tables, see Friberg (2007, tablets MS 3899, VAT 8492, and BM 85200 + VAT 6599 # 23; 56–65).

According to Baqir (2013, 38), these tables were masterfully utilized by the Mesopotamians to solve cubic equations of the form $x^3 + bx^2 = c.$ They began by multiplying the equation through by $\frac{1}{b^3}$ to obtain

$\frac{x^3}{b^3}+\frac{x^2}{b^2}=\frac{c}{b^3}$. Denoting $n=\frac{x}{b}$ and $d=\frac{c}{b^3}$, they got $n^3+n^2=d$, which they solved by searching through their prepared tables.

Tables of Inverse Exponents or Logarithms

Recall that $\log_b x$ is a number y such that $b^y=x$, where b and x are both positive and $b\neq 1$. In words, the logarithm of x is the power y to which the base *b* must be raised to achieve the value of x. Present-day mathematicians work with both common logarithms (base 10) and natural logarithms (base *e*). The constant *e* approximately equals 2.17828 and one of its mathematical definitions is

$$e=\lim_{n\to\infty}\left(1+\frac{1}{n}\right)^n.$$

Tables 4 and 5 show the logarithmic contents of tablet MLC 2078.

Table 4. Tablet MLC 2078: Logarithms to Base 16

Tablet writing	Power interpretation	Logarithmic interpretation, base 16
$\frac{1}{4}$ 2	$16^{\frac{1}{4}} = 2$	$\frac{1}{4} = \log_{16} 2$
$\frac{1}{2}$ 4	$16^{\frac{1}{2}} = 4$	$\frac{1}{2} = \log_{16} 4$
$\frac{3}{4}$ 8	$16^{\frac{3}{4}} = 8$	$\frac{3}{4} = \log_{16} 8$
1 16	$16^{1} = 16$	$1 = \log_{16} 16$
$1\frac{1}{4}$ 32	$16^{1\frac{1}{4}} = 32$	$1\frac{1}{4} = \log_{16} 32$
$1\frac{1}{2}$ 64	$16^{1\frac{1}{2}} = 64$	$1\frac{1}{2} = \log_{16} 64$

Table 5. Tablet MLC 2078: Logarithms to Base 2

Decimal system notation	Power interpretation	Logarithmic interpretation
2 1	$2^{1} = 2$	$1 = \log_2 2$
4 2	$2^{2} = 4$	$2 = \log_2 4$
8 3	$2^{3} = 8$	$3 = \log_2 8$
16 4	$2^{4} = 16$	$4 = \log_2 16$
32 5	$2^{5} = 32$	$5 = \log_2 32$
64 6	$2^{6} = 64$	$6 = \log_2 64$

For further reading on the logarithmic interpretation, see Neugebauer and Sachs (1945, 35).

A Pythagorean Triples Table

Tablet Plimpton 322 was discovered in Larsa, and it dates back to 1800 BCE. Plimpton 322 has been one of the most puzzling Mesopotamian artifact ever discovered. At least one column was broken off the left-hand edge of this rectangular tablet. The surviving part of the tablet is a tabular arrangement comprising fifteen rows and four columns. The wonders of this tablet stem from the fact that it gives the values of fifteen Pythagorean triples (or triplets), which are difficult to generate.

Recall that the Pythagorean triples are three positive integers (*a*, *b*, *c*) such that $a^2 + b^2 = c^2$. Geometrically, this is the statement of the Pythagorean rule: the sum of squares of the two sides (*a* and *b*) of a right triangle is equal to the square of its hypotenuse (*c*). A primitive Pythagorean triple is one in which the triples have no common factor other than one. For example, (3, 4, 5) is a primitive Pythagorean triple, because $3^2 + 4^2 = 5^2$ and the three numbers have no common factor. However, (10, 24, 26) is a Pythagorean triple that is not primitive. We recall here some formulae for generating Pythagorean triples attributed to Greek mathematicians:

Euclid's formula: If *m* and *n* are positive integers with $m < n$, then a Pythagorean triple (a, b, c) can be generated by setting

$$a = n^2 - m^2, \quad b = 2mn, \quad \text{and} \quad c = n^2 + m^2.$$

Example: If $m = 4$ and $n = 6$, then $a = 20$, $b = 48$, and $c = 52$ form a Pythagorean triple that is not primitive.

Euclid's formula generates primitives when *m* and *n* have no common factor, and one of them is even, and the other is odd.

Example: If $m = 4$ and $n = 5$, then $a = 9$, $b = 40$, and $c = 41$ form a Pythagorean triple that is primitive.

Proclus's formula: If n is a positive integer, then a Pythagorean triple (a, b, c) can be generated by setting

$$a = 2n+1, \quad b = 2n(n+1), \quad \text{and} \quad c = 2n(n+1)+1$$

Plimpton 322 essentially is a listing of the first fifteen primitive Pythagorean triples that must have been generated by a certain formula. The tablet demonstrates that the Mesopotamians worked with a rudimentary theory of numbers and had knowledge of Pythagoras theorem hundreds of years before the Greeks. Table 6 is a re-creation of Plimpton 322, where the numbers are in sexagesimal representation (equivalent decimal representations in parentheses.) Column 4, extreme right of Table 6, gives a serial listing from 1 to 15 to the items of the tablet. Column 2 gives the shorter leg (***a***) of a right triangle and column 3 gives the hypotenuse (***c***) of the triangle. Column 1 has been the most difficult to interpret by experts. If a semicolon (the sexagesimal separator) is inserted after the first character in column 1, then according to Neugebauer and Sachs (1945, 38–41), this column gives $\left(\frac{c}{b}\right)^2$ that is, $\sec^2 A$, where A is the acute angle opposite side, a, of the right triangle. Further, the numbers in column 1 are arranged in linearly decreasing order from approximately 1.9834 $= \sec^2 44.76^\circ$ on top to 1.3572$= \sec^2 31.49^\circ$ at the bottom. For further reading and different interpretations of Plimpton 322, see Bruins (1949, 1955), Friberg (2007, 433), Robson (2008, 110–15), and Shekoury (2010, 155–63). For the benefit of the reader, I list the definitions of the basic trigonometric functions:

Let *A* be an angle in a right triangle with opposite sides *a*, adjacent side *b*, and hypotenuse *c*. Then, the basic trigonometric functions are as follows:

$$\sin A = \frac{a}{c} = \frac{opp}{hyp} \quad \cos A = \frac{b}{c} = \frac{adj}{hyp} \quad \tan A = \frac{a}{b} = \frac{opp}{adj}$$

$$\csc A = \frac{c}{a} = \frac{hyp}{opp} \quad \sec A = \frac{c}{b} = \frac{hyp}{adj} \quad \cot A = \frac{b}{a} = \frac{adj}{opp}.$$

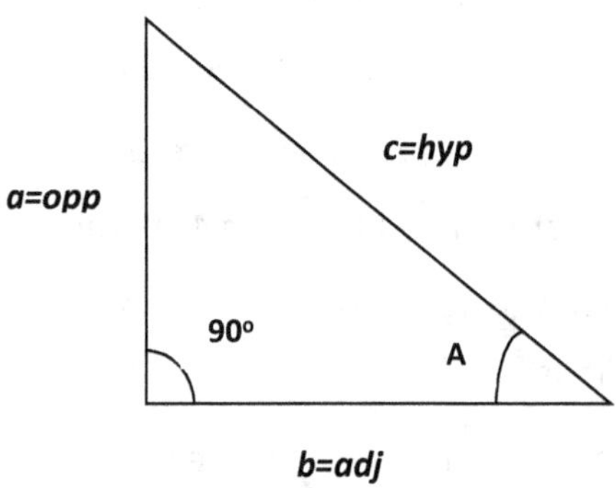

Table 6. Plimpton 322: Pythagorean Triples (Sexagesimal)

Broken off Column	Column 1	Column 2	Column 3	Column 4
	1;59,0,15 (=1.9834)	1,59 (= 119)	2,49 (=169)	1
	1;56,56,58,14,50,6,15	56,7	1,20,25	2
	1;55,7,41,15,33,45	1,16,41	1,50,49	3
	1;53,10,29,32,52,16	3,31,49	5,9,1	4
	1;48,54,1,40	1,5	1,37	5
	1;47,6,41,40	5,19	8,1	6
	1;43,11,56,28,26,40	38,11	59,1	7
	1;41,33,59,3,45	13,19	20,49	8
	1;38,33,36,36	8,1	12,49	9
	1;35,10,2,28,27,24,26, 40	1,22,41	2,16,1	10
	1;33,45	45,0 (45)	1,15,0 (=75)	11
	1;29,21,54,2,15	27,59	48,49	12
	1;27,0,3,45	2,41	4,49	13
	1;25,48,51,35,6,40	29,31	53,49	14
	1;23,13,46,40 (=1.3572)	56 (=56)	1,46 (=116)	15

Arithmetic and Geometric Progressions

To help understand the ensuing discussion, we review some present-day formulae regarding progressions. Recall that an *n*-term geometric progression $a + ar + ar^2 + \cdots + ar^{n-1}$, with first term *a* and ratio $r \neq 1$, sums to $a + ar + ar^2 + \cdots + ar^{n-1} = \frac{a(r^n - 1)}{r - 1}$. (5.1)

Also, recall the following two formulae regarding the sum and the sum of squares of the first n natural numbers (or positive integers):

$$1 + 2 + \cdots + n = \frac{n(n+1)}{2} \quad (5.2)$$

$$1^2 + 2^2 + \cdots + n^2 = \frac{n(n+1)(2n+1)}{6}. \quad (5.3)$$

For a quick review of progressions, see Kohn and Herzog (2011, 229–38). In the ensuing discussions, all numbers are written in the decimal system representation.

Tablet AO 6484 (the Seleucid period, 323–126 BCE) demonstrates that the Mesopotamians used arithmetic and geometric progressions to calculate distributions of money or land among siblings or a group of people. In one case, they calculated the exact sum of $1 + 2 + 4 + \cdots + 2^9$, which is a geometric progression with *n* = 10 terms, first term *a* = 1 and ratio *r* = 2. Applying formula (5.1), a present-day mathematician finds

$$1 + 2 + 4 + \cdots + 2^9 = \frac{1(2^{10} - 1)}{2 - 1} = 2^{10} - 1. \quad (5.4)$$

In tablet AO 6484, a Mesopotamian mathematician calculated

$$1 + 2 + 4 + \cdots + 2^9 = 2^9 + (2^9 - 1). \quad (5.5)$$

The present-day answer in formula (5.4) and the Mesopotamian answer in formula (5.5) are equivalent because

$$2^9 + (2^9 - 1) = 2(2^9) - 1 = 2^{10} - 1.$$

Tablet AO 6484 also contains the calculation of the sum of squares of the first *n* natural numbers using the formula:

$$1^2 + 2^2 + \cdots + n^2 = \left(1 \times \frac{1}{3} + \frac{2}{3} \times n\right)(1 + 2 + \cdots + n). \qquad (5.6)$$

Note that formulae (5.3) and (5.6) are equivalent.
In particular, the Mesopotamian mathematician calculated:

$$\begin{aligned} 1^2 + 2^2 + \cdots + 10^2 &= \left(1 \times \frac{1}{3} + \frac{2}{3} \times 10\right)(1 + 2 + \cdots + 10) \\ &= \left(1 \times \frac{1}{3} + \frac{2}{3} \times 10\right) \times 55 = 385 \end{aligned}$$

For further reading, see Baqir (2013) and Friberg (2007, 323).

CHAPTER 6 A Review of Quadratic Equations

This chapter contains a review of present-day mathematics of quadratic (second-degree) equations in a manner that will help us understand the ancient Mesopotamian mathematics. The reader may consult Kohn and Herzog (2011) for a quick review of algebra and Kohn (2001) for a quick review of geometry.

One of the general forms of writing a quadratic equation is

$$ax^2 + px = q,$$
$$\text{where } a, p, \text{and } q \text{ are numbers such that } a \neq 0. \tag{6.1}$$

When $a = 1$, the quadratic equation reduces to a simpler form,

called the normal or the standard form:

$$x^2 + px = q\,. \tag{6.2}$$

Several methods can be used to solve a quadratic equation. The basic method is that of **completing the square**, which is based on the fact that

$$a^2 + 2ab + b^2 = (a+b)^2\,. \tag{6.3}$$

First, we treat the standard quadratic equation in Formula (6.2); then we will treat the general form in Formula (6.1).

Equations of the Standard Type $x^2 + px = q$

The method of completing the square consists of five steps: ***halving, squaring, completing, extracting,*** **and** ***adjusting***. To solve $x^2 + px = q$ by the method of completing the square, perform the following steps:

Step 1 [halving]: Take half of the coefficient of x, $\frac{1}{2}p$.

Step 2 [squaring]: Multiply $\frac{1}{2}p$ by itself to get $\left(\frac{p}{2}\right)^2$.

Step 3 [completing]: Add $\left(\frac{p}{2}\right)^2$ to both sides of the equation. Then express the left side of the equation as a complete square.

Step 4 [extracting]: Take the square roots of both sides of the equation.

Step 5 [adjusting]: Subtract $\left(\frac{p}{2}\right)$ from both sides of the equation.

In symbolic algebra, the five steps for solving $x^2 + px = q$ are:

$$x^2 + px = q$$

$$\frac{1}{2}p \qquad \text{[halving]}$$

$$\left(\frac{p}{2}\right)^2 \qquad \text{[squaring]}$$

$$x^2 + px + \left(\frac{p}{2}\right)^2 = \left(\frac{p}{2}\right)^2 + q \qquad \text{[completing]}$$

$$\left(x + \frac{p}{2}\right)^2 = \left(\frac{p}{2}\right)^2 + q \qquad \text{[completing]}$$

$$\left(x + \frac{p}{2}\right) = \pm\sqrt{\left(\frac{p}{2}\right)^2 + q} \qquad \text{[extracting]}$$

$$x = \pm\sqrt{\left(\frac{p}{2}\right)^2 + q} \ - \frac{p}{2} \qquad \text{[adjusting]} \qquad (6.4)$$

When $p = 1$, we get $x = \pm\sqrt{\left(\frac{1}{2}\right)^2 + q} \ - \frac{1}{2}$ (6.5)

The same steps of completing the square apply to the quadratic equation $x^2 - px = q$, except that in the adjusting step, we add rather than subtract $\left(\frac{p}{2}\right)$. The resulting formula is:

$$x = \pm\sqrt{\left(\frac{p}{2}\right)^2 + q} \ + \frac{p}{2} \qquad (6.6)$$

When $p = 1$, we get $x = \pm\sqrt{\left(\frac{1}{2}\right)^2 + q} \ + \frac{1}{2}$ (6.7)

Example: Solve $x^2 + 4x = 2$ by the method of completing the square.

Solution: The equation is in the form:

$x^2 + px = q$, where $p = 4$ and $q = 2$.

$x^2 + 4x = 2$	
$\frac{1}{2}(4)$	[halving]
$\left(\frac{4}{2}\right)^2$	[squaring]
$x^2 + 4x + \left(\frac{4}{2}\right)^2 = 2 + \left(\frac{4}{2}\right)^2$	[completing]
$x^2 + 4x + 4 = 2 + 4$	[completing]
$(x+2)^2 = 6$	[completing]
$x + 2 = \pm\sqrt{6}$	[extracting]
$x = -2 \pm \sqrt{6}$	[adjusting]
$x = -2 + \sqrt{6}$ or $x = -2 - \sqrt{6}$.	

Equations of the General Type

$ax^2 + px = q$ or $ax^2 - px = q$

There are two popular methods to solve a quadratic equation of the general form $ax^2 + px = q$: the **reduction to unity** (or standardizing) method and the **transformation method**.

The Reduction to Unity Method

First, divide the equation $ax^2 + px = q$ through by a, the coefficient of x^2. Thus, you obtain the standard form $x^2 + \frac{p}{a}x = \frac{q}{a}$. Then apply the five steps of the method of completing the square as follows:

$$ax^2 + px = q \quad \Rightarrow x^2 + \frac{p}{a}x = \frac{q}{a} \qquad \text{[standardizing]}$$

$$\frac{p}{2a} \qquad \text{[halving]}$$

$$\left(\frac{p}{2a}\right)^2 \qquad \text{[squaring]}$$

$$x^2 + \frac{p}{a}x + \left(\frac{p}{2a}\right)^2 = \left(\frac{p}{2a}\right)^2 + \frac{q}{a} \qquad \text{[completing]}$$

$$\left(x + \frac{p}{2a}\right)^2 = \left(\frac{p}{2a}\right)^2 + \frac{q}{a} \qquad \text{[completing]}$$

$$\left(x + \frac{p}{2a}\right) = \pm\sqrt{\left(\frac{p}{2a}\right)^2 + \frac{q}{a}} \qquad \text{[extracting]}$$

$$x = \pm\sqrt{\left(\frac{p}{2a}\right)^2 + \frac{q}{a}} - \frac{p}{2a} \qquad \text{[adjusting]} \qquad (6.8)$$

Formula (6.8) can be rewritten in another form:

$$x = \frac{\pm\sqrt{\left(\frac{p}{2}\right)^2 + aq} - \frac{p}{2}}{a} \tag{6.9}$$

If $a = 1$, then both formulae (6.8) and (6.9) coincide with Formula (6.4).

Applying the above steps on equations of the form $ax^2 - px = q$, we get:

$$x = \pm\sqrt{\left(\frac{p}{2a}\right)^2 + \frac{q}{a}} + \frac{p}{2a} \quad . \tag{6.10}$$

Or, $x = \pm\frac{\sqrt{\left(\frac{p}{2}\right)^2 + aq} + \frac{p}{2}}{a}$ (6.11)

Example: Solve $2x^2 - 3x = 3$.

$$2x^2 - 3x = 3$$

$$x^2 - \frac{3}{2}x = \frac{3}{2} \qquad \text{[standardizing]}$$

$$\frac{1}{2}\left(-\frac{3}{2}\right) = -\frac{3}{4} \qquad \text{[halving]}$$

$$\left(-\frac{3}{4}\right)^2 = \frac{9}{16} \qquad \text{[squaring]}$$

$$x^2 - \frac{3}{2}x + \frac{9}{16} = \frac{3}{2} + \frac{9}{16} \qquad \text{[completing]}$$

$$\left(x - \frac{3}{4}\right)^2 = \frac{33}{16} \qquad \text{[completing]}$$

$$\left(x - \frac{3}{4}\right) = \pm\sqrt{\frac{33}{16}} \qquad \text{[extracting]}$$

$$x = \frac{3}{4} \pm \frac{1}{4}\sqrt{33} \qquad \text{[adjusting]}$$

$$x = \frac{1}{4}\left(3 + \sqrt{33}\right) \quad \text{or} \quad x = \frac{1}{4}\left(3 - \sqrt{33}\right)$$

The Transformation Method

To solve a quadratic equation of the form $ax^2 + px = q$ by **transformation**, perform the following steps:

Multiply $ax^2 + px = q$ through by a, the coefficient of x^2, to obtain $a^2x^2 + pax = aq$.

Set $y = ax$, to obtain the transformed equation $y^2 + py = aq$, which can be solved using Formula (6.4). Thus

$$y = \pm\sqrt{\left(\frac{p}{2}\right)^2 + aq} \;\; - \frac{p}{2}$$

$$ax = \pm\sqrt{\left(\frac{p}{2}\right)^2 + aq} \;\; - \frac{p}{2}$$

$$x = \frac{\pm\sqrt{\left(\frac{p}{2}\right)^2 + aq} \;\; - \frac{p}{2}}{a}$$

, which is same as the reduction to unity Formula (6.9).

Applying similar steps, we can solve $ax^2 - px = q$ by transformation to get

$$x = \frac{\pm\sqrt{\left(\frac{p}{2}\right)^2 + aq} \;\; + \frac{p}{2}}{a}$$

The Present-day Quadratic-Formula Method

Most present-day algebra books write the general quadratic equation in the form:

$ax^2 + bx + c = 0$, where a, b, and c are any real numbers such that $a \neq 0$.

The general solution is given by the well-known **quadratic formula:**

$$x = \frac{-b \pm \sqrt{b^2 - 4ac}}{2a}$$

The **quadratic formula** can be derived from the method of completing the square as in the following steps:

$$ax^2 + bx + c = 0$$

$$x^2 + \frac{b}{a}x + \frac{c}{a} = 0 \qquad \text{[standardi zing]}$$

$$x^2 + \frac{b}{a}x = -\frac{c}{a}$$

$$\frac{1}{2}\left(\frac{b}{a}\right) = \frac{b}{2a} \qquad \text{[halving]}$$

$$\left(\frac{\text{b}}{2\text{a}}\right)^2 \qquad \text{[squaring]}$$

$$x^2 + \frac{b}{a}x + \left(\frac{b}{2a}\right)^2 = \left(\frac{b}{2a}\right)^2 - \frac{c}{a} \qquad \text{[completing]}$$

$$\left(x + \frac{b}{2a}\right)^2 = \left(\frac{b}{2a}\right)^2 - \frac{c}{a} \qquad \text{[completing]}$$

$$\left(x + \frac{b}{2a}\right) = \pm\sqrt{\left(\frac{b}{2a}\right)^2 - \frac{c}{a}} \qquad \text{[extracting]}$$

$$x = \pm\sqrt{\left(\frac{b}{2a}\right)^2 - \frac{c}{a}} \;\; - \frac{b}{2a} \qquad \text{[adjusting]}$$

$$x = \pm\frac{1}{2a}\sqrt{\text{b}^2 - 4ac} \;\; - \frac{b}{2a}$$

$$x = \frac{-b \pm \sqrt{b^2 - 4ac}}{2a} \qquad (6.12)$$

Formula (6.12) is the most general method for solving quadratic equations in present-day textbooks. Note that:

The sum of the roots $= -\dfrac{b}{a}$ and the product of the roots $= \dfrac{c}{a}$

To complete the review of quadratic equations, we discuss the nature of the roots of the equation. There are four types of possible solutions (roots or zeros) of a quadratic equation: **two distinct real rational roots; two distinct real irrational roots; two equal (or double) real roots; or no real roots but two conjugate complex (imaginary) roots.** The expression $D = b^2 - 4ac$, called the **discriminant**, determines the nature of the roots of the quadratic equation according to the following four scenarios:

1. If D is positive and is a perfect square, then the equation has two distinct real, rational roots.
2. If D is positive and is not a perfect square, then the equation has two distinct irrational, real roots.
3. If D is zero, then the equation has two equal (or double) real roots.
4. If D is negative, then the equation has no real root but two complex (imaginary) roots.

Example: Solve $2x^2 - 3x - 3 = 0$.

$$2x^2 - 3x - 3 = 0 \quad \Rightarrow x = \frac{-(-3) \pm \sqrt{(-3)^2 - 4(2)(-3)}}{2(2)}$$

$$x = \frac{3 \pm \sqrt{9+24}}{4} \quad \Rightarrow \quad x = \frac{3 \pm \sqrt{33}}{4} \quad \Rightarrow \quad x = \frac{1}{4}\left(3 \pm \sqrt{33}\right)$$

$$x = \frac{1}{4}\left(3 + \sqrt{33}\right) \quad \text{or} \quad x = \frac{1}{4}\left(3 - \sqrt{33}\right)$$

Note: The discriminat $D = b^2 - 4ac = (-3)^2 - 4(2)(-3) = 33$ is postive but is not a perfectsquare. Thefore, the resulting roots were two distinct irrational real numbers.

Example: Solve $2x^2 - 4x + 3 = 0$.

$$2x^2 - 4x + 3 = 0 \quad \Rightarrow x = \frac{-(-4) \pm \sqrt{(-4)^2 - 4(2)(3)}}{2(2)}$$

$$x = \frac{4 \pm \sqrt{16 - 24}}{4} \quad \Rightarrow \quad x = \frac{4 \pm \sqrt{-8}}{4} \quad \Rightarrow \quad x = \frac{1}{4}\left(1 \pm \sqrt{-8}\right)$$

$$x = \frac{1}{4}\left(1 \pm \sqrt{(-1)(8)}\right) \quad \Rightarrow \quad x = \frac{1}{4}\left(1 \pm \sqrt{-1}\,\sqrt{8}\right)$$

$$x = \frac{1}{4}\left(1 \pm i2\sqrt{2}\right) \quad \Rightarrow x = \frac{1}{4}\left(1 \pm i\sqrt{2}\right)$$

$$x = \frac{1}{4}\left(1 + i\sqrt{2}\right) \quad \text{or} \quad x = \frac{1}{4}\left(1 - i\sqrt{2}\right), \text{where } i = \sqrt{-1}.$$

Note: The discriminat $D = b^2 - 4ac = (-4)^2 - 4(2)(3) = -8$ is negative. Thefore, the rootswere complex numbers.

Other Quick Methods for Solving Quadratic Equations

First, recall the Zero-Product Rule (or Zero Property):

if $a \cdot b = 0$, then either $a = 0$ or $b = 0$, or both.

The Factoring Method: First, collect all terms on one side of the equation and keep zero on the other side. Second, factor out the quadratic expression. Third, equate each factor to zero and solve.

Example: Solve $3x^2 - 10x = 8$

$$3x^2 - 10x = 8 \quad \Rightarrow \quad 3x^2 - 10x - 8 = 0 \quad \Rightarrow (3x + 2)(x - 4) = 0$$

$$(3x + 2) = 0 \quad \text{or} \quad (x - 4) = 0 \quad \Rightarrow x = -2/3 \text{ or } x = 4.$$

Note: The factoring method does not always work because some quadratic expressions are not factorable. For example, $x^2 + 4x - 2 = 0$, cannot be factored.

Warning: If you divide an equation by the unknown variable, you lose one of the two solutions.

Example: Solve $x^2 - 7x = 0$.

Solution1 : Divide both sides by x to get
$x - 7 = 0 \quad \Rightarrow x = 7$, only one solution.
The second solution $x = 0$ has been lost.

Solution 2 : $x^2 - 7x = 0 \quad \Rightarrow \quad x(x-7) = 0$

$$x = 0 \quad \text{or} \quad (x-7) = 0.$$

The two solutions are : $x = 0$ and $x = 7$.
No solution has been lost.

Reducing Higher-Order Equations to Quadratic Form

Some higher-order equations can be reduced to the quadratic form $ax^2 + bx + c = 0$ and then be solved by factoring, completing the square, or by the general quadratic formula. We give an example of this reduction method because the Mesopotamian mathematicians did work with such higher-order equations.

Example (Kohn and Herzog 2011, p 169)**:** Solve the fourth-degree equation $x^4 - 13x^2 + 36 = 0$.

Setting $y = x^2$, the equation becomes $y^2 - 13y + 36 = 0$.

Now apply the quadratic formula (6.12):

$$y = \frac{-(-13) \pm \sqrt{(-13)^2 - 4(1)(36)}}{2(1)}$$

$$y = \frac{13 \pm \sqrt{169 - 144}}{2} \quad \Rightarrow \quad y = \frac{13 \pm \sqrt{25}}{2}$$

$$y = \frac{13 \pm 5}{2} \quad \Rightarrow \quad y = 9 \quad \text{or} \quad y = 4.$$

Recalling that $y = x^2$, we get $x = \pm\sqrt{y}$

$$x = \pm\sqrt{9} = \pm 3 \quad \text{or} \quad x = \pm\sqrt{4} = \pm 2.$$

The four solutions are : $x = -3, -2, 2, 3$.

Example: Solve $x - 5\sqrt{x} + 6 = 0$.

Setting $y = \sqrt{x}$, the equation becomes $y^2 - 5y + 6 = 0$.
Now apply the quadratic formula:

$$y = \frac{-(-5) \pm \sqrt{(-5)^2 - 4(1)(6)}}{2(1)} \quad \Rightarrow y = \frac{5 \pm \sqrt{25 - 24}}{2}$$

$$y = \frac{5 \pm \sqrt{1}}{2} \quad \Rightarrow y = \frac{5 \pm 1}{2} \quad \Rightarrow \quad y = 3 \quad \text{or} \quad y = 2.$$

Recalling that $y = \sqrt{x}$, we get $x = y^2$.

Therefore, $x = (3)^2 = 9 \quad \text{or} \quad x = (2)^2 = 4$.

Example: Solve $\sqrt{x^2 - 10x} - 1 = 0$.

First, isolate the radical expression on one side.

$$\sqrt{x^2 - 10x} = 1 \quad \Rightarrow \left(\sqrt{x^2 - 10x}\right)^2 = (1)^2 \Rightarrow x^2 - 10x = 1$$

$x^2 - 10x - 1 = 0$, which is in a quadratic form.

Now apply the quadratic formula (6.12):

$$x = \frac{-(-10) \pm \sqrt{(-10)^2 - 4(1)(-1)}}{2(1)} = \frac{10 \pm \sqrt{100+4}}{2}$$

$$x = \frac{10 \pm \sqrt{104}}{2} = \frac{10 \pm \sqrt{2 \times 4 \times 13}}{2} = \frac{10 \pm 2\sqrt{2}\sqrt{13}}{2}$$

$x = 5 \pm \sqrt{2}\sqrt{13}$.

CHAPTER 7 Problem Texts: Algebraic Equations

The discussions in this chapter and the previous chapters present concrete evidence that the Mesopotamians invented masterful methods to solve algebraic problems comparable to our present-day high-school mathematics. However, their methods did not employ symbolic algebra as we know it today. The Mesopotamians employed mathematical procedures that include completing the square, false position, auxiliary variable, reduction, and substitution schemes. Words and idioms, rather than symbols, were used to express addition, subtraction, multiplication, reciprocation, roots, and powers of numbers. This chapter discusses some Mesopotamian cuneiform tablets that deal with quadratic and higher-degree mathematical equations.

Quadratic (Second-Degree) Equations

The Mesopotamian mathematicians handled most kinds of second-degree equations, except those leading to negative or imaginary roots (solutions). In this section, we present some cuneiform tablets that deal with the following types of quadratic equations (where *a*, *p*, and *q* are positive numbers):

Type 1: $x^2 + x = q$, **Type 2:** $x^2 - x = q$, **Type 3:** $x^2 + px = q$

Type 4: $x^2 - px = q$, **Type 5:** $ax^2 + x = q$, **Type 6:** $ax^2 + px = q$

To facilitate understanding the ancient mathematical texts, we adopt the following two conventions:

Convention 1: First, we give an English translation of the tablet contents. A typical Mesopotamian mathematical tablet contains a data part, a question part, and a step-by-step worked-out solution part. Second, we give the present-day symbolic algebraic interpretation of the tablet contents. The numbers that appear in the translation are in decimal form, and their sexagesimal equivalents appear in parentheses ( ).

Convention 2: At each step in the translated solution, we insert contents, in brackets [ ], that link the Mesopotamian solution to the steps of the method of completing the square. As we discussed in chapter 6, the steps of the method of completing the square are: **halving, squaring, completing, extracting, and adjusting.**

Tablet BM 13901 # 1. Equation Type 1: $x^2 + x = q$

Translation:

The side was added to the area of a square giving $\frac{3}{4}$ (= 0;45). What is the side length? Take half of 1 [halving], which is $\frac{1}{2}$ (= 0;30). Multiply the result by itself [squaring] to get $\frac{1}{4}$ (= 0;15). Add $\frac{1}{4}$ to $\frac{3}{4}$ to get 1 [completing]. Take the square root of 1 to obtain 1 [extracting]. Subtract from one the half [adjusting], which you squared, to obtain $\frac{1}{2}$. That is the side length of the square.

Present-day symbolic mathematics: tablet BM 13901 # 1

If we let x = side of the square, then the equation to be solved is $x^2 + x = \frac{3}{4}$, which is of Type 1: $x^2 + x = q$.

$\frac{1}{2}(1) = \frac{1}{2}$ [halving]

$\frac{1}{2} \times \frac{1}{2} = \frac{1}{4}$ [squaring]

$x^2 + x + \frac{1}{4} = \frac{3}{4} + \frac{1}{4}$ [completing]

$\left(x + \frac{1}{2}\right)^2 = 1$ [completing]

$x + \frac{1}{2} = 1$ [extracting]

$x = 1 - \frac{1}{2} = \frac{1}{2}$ [adjusting]

Note that when extracting square roots, we disregarded the negative root because the Mesopotamians did not work with negative square roots.

Tablet BM 13901 # 2. Equation Type 2: $x^2 - x = q$

Translation:

The side was subtracted from the area of a square giving 870 (= 14,30). What is the side? Take half of 1, which is $\frac{1}{2}$ (= 0;30) [halving]. Multiply the result by itself to get $\frac{1}{4}$ (= 0;15) [squaring]. Add $\frac{1}{4}$ to 870 to get $870\frac{1}{4}$ (= 14,30;15) [completing]. Take the square root of $870\frac{1}{4}$ to obtain $29\frac{1}{2}$

(= 29;30) [extracting]. Add the half [adjusting], which you squared, to $29\frac{1}{2}$, you obtain 30. That is the side of the square.

Present-day symbolic mathematics: tablet BM 13901 # 2

If we let $x =$ side of the sqaure, then the equation to be solved is $x^2 - x = 870$, which is of Type $2: x^2 - x = q$.

$$\frac{1}{2}(1) = \frac{1}{2} \qquad \text{[halving]}$$

$$\frac{1}{2} \times \frac{1}{2} = \frac{1}{4} \qquad \text{[squaring]}$$

$$x^2 - x + \frac{1}{4} = 870 + \frac{1}{4} \qquad \text{[completing]}$$

$$\left(x - \frac{1}{2}\right)^2 = 870\frac{1}{4} \qquad \text{[completing]}$$

$$x - \frac{1}{2} = \sqrt{870\frac{1}{4}} = 29\frac{1}{2} \qquad \text{[extracting]}$$

$$x = 29\frac{1}{2} + \frac{1}{2} = 30 \qquad \text{[adjusting]}$$

Tablet BM 13901 # 5. Equation Type 3: $x^2 + px = q$.

Translation:

You added to the area of a square one and one-third of its side. The result was $\frac{11}{12}$ (= 0;55). What is the side length? Take unity and add to it one-third of unity to get the result $1\frac{1}{3}$ (= 1;20). Halve this result to get $\frac{2}{3}$ (= 0;40) [halving]. Square $\frac{2}{3}$ to get $\frac{4}{9}$ (= 0;26,40) [squaring]. Add $\frac{4}{9}$ to $\frac{11}{12}$ to get

$\frac{49}{36}$ (= 1;20,40) [completing]. Take the square root of $\frac{49}{36}$ to get $\frac{7}{6}$ (= 1;10) [extracting]. Subtract $\frac{2}{3}$ which you squared from $\frac{7}{6}$ to get $\frac{1}{2}$ (= 0;30) [adjusting]. That is the side length of the square.

Present-day symbolic mathematics: tablet BM 13901 # 5

If we let $x =$ side of the square, then the equation to be solved is

$x^2 + x + \frac{1}{3}x = \frac{11}{12}$ or $x^2 + 1\frac{1}{3}x = \frac{11}{12}$; of Type 3 : $x^2 + px = q$.

$\frac{1}{2}(1\frac{1}{3}) = \frac{2}{3}$ [halving]

$\frac{2}{3} \times \frac{2}{3} = \frac{4}{9}$ [squaring]

$x^2 + 1\frac{1}{3}x + \frac{4}{9} = \frac{11}{12} + \frac{4}{9}$ [completing]

$\left(x + \frac{2}{3}\right)^2 = \frac{49}{36}$ [completing]

$x + \frac{2}{3} = \sqrt{\frac{49}{36}}$ [extracting]

$x = \frac{7}{6} - \frac{2}{3} = \frac{1}{2}$. [adjusting]

The Mesopotamians used Formula (6.4): $x = \sqrt{\left(\frac{p}{2}\right)^2 + q} - \frac{p}{2}$.

Tablet YBC 6967. Equation Type 4: $x^2 - px = q$

Translation:

A number exceeds its reciprocal by 7. What are the number and the reciprocal? You need to halve 7, the amount by which the number exceeds its reciprocal to get $3\frac{1}{2}$ [halving]. Multiply $3\frac{1}{2}$ by $3\frac{1}{2}$ to get $12\frac{1}{4}$ [squaring]. Add your result of $12\frac{1}{4}$ to 60 (which is the multiplication of the number by its reciprocal) to get $72\frac{1}{4}$ [completing]. Take the square root of $72\frac{1}{4}$ to get $8\frac{1}{2}$ [extracting]. Subtract $8\frac{1}{2} - 3\frac{1}{2}$ and then add $8\frac{1}{2} + 3\frac{1}{2}$. In the first case you get 5, and in the second case you get 12 [adjusting]. This gives the number and its reciprocal.

Present-day symbolic mathematics: tablet YBC 6967

If we denote the number and its reciprocal by x and y, respectively, then the problem reduces to two equations: $x - y = 7$ and $xy = 60$. Through substitution, the two equations lead to one quadratic equation $x^2 - 7x = 60$, which is of Type 4: $x^2 - px = q$.

$\frac{1}{2}(7) = 3\frac{1}{2}$ [halving]

$3\frac{1}{2} \times 3\frac{1}{2} = 12\frac{1}{4} = \frac{49}{4}$ [squaring]

$x^2 - 7x + \frac{49}{4} = 60 + \frac{49}{4} = 72\frac{1}{4}$ [completing]

$\left(x - \frac{7}{2}\right)^2 = 72\frac{1}{4}$ [completing]

$x - \frac{7}{2} = \sqrt{72\frac{1}{4}} = 8\frac{1}{2}$ [extracting]

$x = 8\frac{1}{2} + \frac{7}{2}$ [adjusting]

$x = 12$

$y = x - 7 = 12 - 7 = 5$

In effect, the Mesopotamians used Formula (6.6)

$$x = \sqrt{\left(\frac{p}{2}\right)^2 + q} + \frac{p}{2}.$$

Non-standard Quadratic Equations

Now we look at more general quadratic equations where the leading coefficient is not 1. These equations are of

Type 5: $ax^2 + x = q$ and Type 6: $ax^2 + px = q$.

As was discussed in chapter 5, the general quadratic equation can be solved by one of two methods: the reduction to unity method or the transformation method.

The Mesopotamian mathematicians used the transformation method to a larger extent than the reduction to unity method. The transformation was later used by the Greek mathematician Diophantus (third century BCE). However, the Muslim mathematician al-Khwarizmi (780 – 850 CE) used the reduction to unity method.

As was discussed in chapter 6, the transformation method for solving an equation of Type 6: $ax^2 + px = q$ leads to using Formula (6.9):

$$x = \frac{\sqrt{\left(\frac{p}{2}\right)^2 + aq} - \frac{p}{2}}{a}.$$

As a special case of Type 6, Type 5: $ax^2 + x = q$, is solved through

$$x = \frac{\sqrt{\left(\frac{1}{2}\right)^2 + aq} - \frac{1}{2}}{a}.$$

Note that we showed here only the (+) sign of the square root because the Mesopotamians did not work with negative parts of the root.

In the text below, we will see that the Mesopotamians applied the steps of the method of transformation as detailed in chapter 6.

Tablet BM 13901 # 4. Equation Type 5: $ax^2 + x = q$

Translation:

You subtracted from the area of a square one-third of itself and added to it the side length, which resulted in $286\frac{2}{3}$ (= 4,46;40). What is the side length? Take unity and subtract from it one-third of itself to get the result $\frac{2}{3}$ (= 0;40). Multiply $\frac{2}{3}$ by $286\frac{2}{3}$ to get $191\frac{1}{9}$ (= 3,11;6,40) [transforming]. Take half unit [halving] and square it to get $\frac{1}{4}$ [squaring], add the quarter to $191\frac{1}{9}$ to get $191\frac{13}{36}$ [completing]. Take its square root to get $13\frac{5}{6}$ [extracting]. Subtract the half that you squared to obtain $13\frac{1}{3}$ (= 13;20) [adjusting]. Multiply the reciprocal of $\frac{2}{3}$ which is $1\frac{1}{2}$ (= 1;30) by $13\frac{5}{6}$ to get 20 [adjusting]. The side length of the square is 20.

Present - day symbolic mathematics of tablet BM 13901 # 4 :

Let x and x^2 denote the side and the area of the square. The Mesopotamian performed the following steps of transforming and completing the square : You subtracted from the area of a square one third of itself means :

$x^2 - \frac{1}{3}x^2 = \frac{2}{3}x^2$. The equation to be solved is $\frac{2}{3}x^2 + x = 286\frac{2}{3}$.

$$\left(\frac{2}{3}\right)\left(\frac{2}{3}\right)x^2 + \left(\frac{2}{3}\right)x = 286\frac{2}{3} \times \frac{2}{3}$$ [transforming]

$$\left(\frac{2}{3}x\right)^2 + \left(\frac{2}{3}x\right) = 191\frac{1}{9}$$ [transforming]

$$y^2 + y = 191\frac{1}{9}, \text{ where } y = \frac{2}{3}x$$ [transforming]

Then consider $\left(\frac{2}{3}x\right)$ as one entity with coefficient equals 1 :

$$\frac{1}{2}(1) = \frac{1}{2}$$ [halving]

$$\left(\frac{1}{2}\right)^2 = \frac{1}{4}$$ [squaring]

$$\left(\frac{2}{3}x\right)^2 + \left(\frac{2}{3}x\right) + \frac{1}{4} = 191\frac{1}{9} + \frac{1}{4}$$ [completing]

$$\left(\frac{2}{3}x + \frac{1}{2}\right)^2 = 191\frac{13}{36}$$ [completing]

$$\frac{2}{3}x + \frac{1}{2} = \sqrt{191\frac{13}{36}} = 13\frac{5}{6}$$ [extracting]

$$\frac{2}{3}x = 13\frac{5}{6} - \frac{1}{2} = 13\frac{1}{3}$$ [adjusting]

$$x = \frac{3}{2}\left(13\frac{1}{3}\right) = 20.$$ [adjusting]

Tablet BM 13901 # 3. Equation Type 6: $ax^2 + px = q$

Translation:

You subtracted from the area of a square one-third of itself and added to it one-third of the side, which resulted in $\frac{1}{3}$(0;20). What is the side length of the square? Take unity and subtract from it one-third. Multiply the remainder $\frac{2}{3}$ (= 0;40) by $\frac{1}{3}$ to get $\frac{2}{9}$ (= 0;13,20) [transforming], hold it. Take half of the one-third which you subtracted [halving], and square the result $\frac{1}{6}$ (= 0;10) to get $\frac{1}{36}$ (= 0;1,40) [squaring]. Add $\frac{1}{36}$ to $\frac{2}{9}$ to get $\frac{1}{4}$ (= 0;15) [completing]. Take the square root of $\frac{1}{4}$ to get $\frac{1}{2}$ [extracting]. Subtract the $\frac{1}{6}$ which you squared from $\frac{1}{2}$ to obtain $\frac{1}{3}$ [adjusting]. Multiply the reciprocal of $\frac{2}{3}$ which is $\frac{3}{2}$ (= 0;1,30) by $\frac{2}{3}$ to get $\frac{1}{2}$ [adjusting]. That is the side length of the square.

Present-day symbolic mathematics: tablet BM 13901 # 3

Let x and x^2 denote the side and area of the square. The Mesopotamian mathematician performed the following steps of transforming and completing the square method:
You subtracted from the area of a square one third of itself means :
$x^2 - \frac{1}{3}x^2 = \frac{2}{3}x^2$.
The equation to be solved is $\frac{2}{3}x^2 + \frac{1}{3}x = \frac{1}{3}$, which is of Type 6.

$\left(\frac{2}{3}\right)\left(\frac{2}{3}\right)x^2 + \left(\frac{1}{3}\right)\left(\frac{2}{3}\right)x = \frac{1}{3} \times \frac{2}{3}$ [transforming]

$\left(\frac{2}{3}x\right)^2 + \left(\frac{1}{3}\right)\left(\frac{2}{3}x\right) = \frac{2}{9}$ [transforming]

$y^2 + \frac{1}{3}y = \frac{2}{9}$, where $y = \frac{2}{3}x$ [transforming]

The Mesopotamian mathematician applied the method of completing the square by considering $\left(\frac{2}{3}x\right)$ as one entity with coefficient equals 1:

Equation : $\left(\frac{2}{3}x\right)^2 + \left(\frac{1}{3}\right)\left(\frac{2}{3}x\right) = \frac{2}{9}$

$\frac{1}{2}\left(\frac{1}{3}\right) = \frac{1}{6}$ [halving]

$\left(\frac{1}{6}\right)^2 = \frac{1}{36}$ [squaring]

$\left(\frac{2}{3}x\right)^2 + \left(\frac{2}{3}x\right) + \frac{1}{36} = \frac{2}{9} + \frac{1}{36}$ [completing]

$\left(\frac{2}{3}x + \frac{1}{6}\right)^2 = \frac{1}{4}$ [completing]

$\frac{2}{3}x + \frac{1}{6} = \sqrt{\frac{1}{4}} = \frac{1}{2}$ [extracting]

$\frac{2}{3}x = \frac{1}{2} - \frac{1}{6} = \frac{1}{3}$ [adjusting]

$x = \frac{3}{2}\left(\frac{1}{3}\right) = \frac{1}{2}$. [adjusting]

Merzbach and Boyer (2011, 29) state that in one ancient text the equation $11x^2 + 7x = 6\frac{1}{4}$ was transformed by the Babylonians to the standard form by multiplying through by 11. The resulting equation, $(11x)^2 + 7(11x) = 68\frac{3}{4}$, can then be written as $y^2 + 7y = 68\,\frac{3}{4}$,

where $y = 11x$. Formula (6.4) gives the solution: $y = \sqrt{\left(\frac{p}{2}\right)^2 + q} - \frac{p}{2}$,

where $p = 7$ and $q = 68\frac{3}{4}$.

Merzbach and Boyer (2011, 29) add: "This solution is remarkable an instance of the use of algebraic transformations."

Reducing Higher-Order Equations to Quadratics

The Mesopotamians handled some higher-degree equations by reducing them to quadratics first. Neugebauer (1969, 47–48) discusses the following cuneiform tablet from Susa (second millennium BCE) that deals with an eighth-degree equation.

Eighth-Degree Equation: A rectangle has length *x*, width y ,and diagonal z subject to the conditions:

$xy = 20{,}0_{60}$ and $x^3z = 14{,}48{,}53{,}20_{60}$, in the sexagesimal system.

What are the length and width of the rectangle?

Solution: In decimal representation, the conditions of the problem are:

$xy = 1200$ and $x^3 z = 3200000$. To simplify the steps of the solution, let

$c_1 = 1200$ and $c_2 = 3200000$. The conditions become

$xy = c_1$ and $x^3 z = c_2$. Thus, $y = \frac{c_1}{x}$ and $z = \frac{c_2}{x^3}$.

The Pythagorean rule leads to:

$$x^2 + y^2 = z^2 \quad \Rightarrow x^2 + \frac{c_1^2}{x^2} = \left(\frac{c_2}{x^3}\right)^2$$

$$x^2 + \frac{c_1^2}{x^2} = \frac{c_2^2}{x^6} \quad \Rightarrow x^8 + c_1^2 x^4 = c_2^2, \text{ which is an eighth - degree equaion.}$$

The Mesopotamian mathematician reduced this eighth-degree equation through the transformation (or substitution) $w = x^4$. This leads to the quadratic equation:

$$w^2 + pw = q, \text{ where } p = c_1^2 \text{ and } q = c_2^2.$$

Applying the method of completing the square, the Mesopotamian mathematician found:

$$w = 2560000_{10} = 11{,}51{,}6{,}40_{60}.$$

Therefore, $x = 40$ and $y = 30$ are the length and the width.

Cubic Equations

Some excavated cuneiform tablets demonstrate that the Mesopotamian mathematicians worked with cubic equations of the pure form $x^3 = c$ and the mixed form $x^3 + x^2 = c$.

Pure cubic (one-term cubic): $x^3 = c$

They solved equations of the form $x^3 = c$ through direct reference to their prepared cube-root tables as given below:

1	1 BA-SI	meaning $\sqrt[3]{1} = 1$
8	2 BA-SI	meaning $\sqrt[3]{8} = 2$
27	3 BA-SI	meaning $\sqrt[3]{27} = 3$

Example: $x^3 = 0;7,30$ (in the sexagesimal system) was solved by interpolating in cube-root tables to read off the cube root of 0;7,30 as $x = 0;30_{60}$. In decimal representation, we have:

$$0;7,30_{60} = \frac{7}{60} + \frac{30}{60^2} = \frac{7}{60} + \frac{30}{3600} = \frac{420+30}{3600} = 0.125_{10}$$

Thus, $x^3 = 0;7,30$ becomes $x^3 = 0.125$ in the decimal system.

$$x^3 = 0.125 \quad \Rightarrow x = \sqrt[3]{0.125} \quad \Rightarrow x = 0.50_{10}$$

Of course, $x = 0;30_{60}$ is equivalent to $x = 0.50_{10}$.

Mixed Cubic-quadratic of Standard (Normal) Form: $x^3 + x^2 = c$

The Mesopotamians solved cubic equations of the form $x^3 + x^2 = c$ by direct reference to tables that listed values of the combination $n^3 + n^2$ of integral values of *n* from 1 to 30. Examples: Tablets MS 3899+VAT 8492 and BM 85200+VAT 6599.

2 1 BA-SI meaning $1^2 + 1^3 = 2$

12 2 BA-SI meaning $2^2 + 2^3 = 12$

36 3 BA-SI meaning $3^2 + 3^3 = 36$

80 4 BA-SI meaning $4^2 + 4^3 = 80$

With the aid of these tables, the Mesopotamians easily read off 6 as the solution for $x^3 + x^2 = 4{,}12_{60}$.

More General Cubics: $ax^3 + bx^2 = c$

The Mesopotamians multiplied the equation $ax^3 + bx^2 = c$ through by $\left(\dfrac{a^2}{b^3}\right)$ to obtain the transformed equation

$\left(\dfrac{ax}{b}\right)^3 + \left(\dfrac{ax}{b}\right)^2 = \dfrac{ca^2}{b^3}$. This is a cubic equation of the standard type in the unknown $y = \left(\dfrac{ax}{b}\right)$. Reading the value of *y* off the already prepared tables, the value of *x* can be determined.

Example: Merzbach and Boyer (2011, 31)

Solve $144x^3 + 12x^2 = 21$

The Mesopotamians used the method of substitution as follows:

Multiplying through by 12 and setting $y = 12x$, the equation becomes

$y^3 + y^2 = 252$. Reading off the mixed cubic-quadratic tables, we get *y* = 6. Thus, $x = \frac{y}{12} = \frac{1}{2}$.

Merzbach and Boyer (2011, 31) state: "Whether the Babylonians were able to reduce the general four-term cubic, $ax^3 + bx^2 + cx = d$, to the normal form is not known."

Few other cases leading to third-degree equations of various types were inscribed in the tablet BM 85200 + VAT 6599. According to Hoyrup (2002, 137–62), this tablet is broken; one part is preserved in the British Museum while the other is in the Berlin collection of Vorderasiatische Texte.

Tablet BM 85200 + VAT 6599 # 5

Translation:

A rectangular trench (excavation or cellar) has its width equal to two-thirds of its length. The length equals to the depth. A volume of soil (or mud) was taken out and appended to the base to come up with $1\frac{1}{6}$ (= 1;10). What are the length and width? Take the reciprocal of $\frac{1}{3}$ (= 0;20) to come up with 3 (= 3). Halve 3 to come up with $1\frac{1}{2}$ (= 1;30), which is the ratio of the length. Take the reciprocal of $1\frac{1}{2}$ (= 1;30) to come up with $\frac{2}{3}$ (= 0;40), which is the ratio of the width. Take the reciprocal of 12 (= 12) which is the ratio of the depth to come up with $\frac{1}{12}$ (= 0;5) and multiply it by 1 to come up with $\frac{1}{12}$ (= 0;5). Again multiply $\frac{1}{12}$ by $\frac{2}{3}$ (= 0;40) to come up with $\frac{1}{18}$ (= 0;3,20). Multiply $\frac{1}{18}$ (= 0;3,20) by $\frac{1}{12}$ (= 0;5) to come up with $\frac{1}{216}$ (= 0;0,16;40). Take the reciprocal of $\frac{1}{216}$ to come up with 216 (= 3,36). Multiply 216 by $1\frac{1}{6}$ (= 1;10) to come up with 252 (= 4,12), which is the sum of the cube and square of number 6 (=6). Multiply 6 by $\frac{1}{12}$ (= 0;5) to come up with $\frac{1}{2}$ (= 0;30), which the length. Multiply 6 by $\frac{1}{18}$ (= 0;3,20) to come up with $\frac{1}{3}$ (= 0;20), which is the width. Multiply 6 by 1 to come up with 6, which is the depth. This is the solution.

Present-day symbolic mathematics: tablet BM 85200 + VAT 6599 # 5:

Mathematically, this case is about a parallelogram whose length, width and height (depth) we denote by x, y, z, respectively. The algebraic setup of the problem is represented by the following three equations:

$$xyz + xy = 1\frac{1}{6}, \quad y = \frac{2}{3}x, \quad and\ z = 12x .$$

Note: The Mesopotamian mathematicians measured vertical dimensions (depth) by units of 12 times the units of horizontal dimensions (length and width); thus, the equation $z = 12x$ arose.

By substitution, the system yields

$$x \times \frac{2}{3} x \times 12x + \frac{2}{3} x^2 = 1\frac{1}{6} \quad \Rightarrow \quad \frac{2}{3} \times 12x^3 + \frac{2}{3} x^2 = 1\frac{1}{6}.$$

The Mesopotamian mathematician multiplied both sides of the last equation by $\frac{3}{2} \times 12^2$ to come up with

$$12^3 x^3 + 12^2 x^2 = 1\frac{1}{6} \times \frac{3}{2} \times 12^2 \quad \Rightarrow \quad 12^3 x^3 + 12^2 x^2 = 252$$

$$(12x)^3 + (12x)^2 = 252.$$

Using their prepared tables for $n^3 + n^2 = c$, the sum of cubes and squares of numbers, they found that 252 is the sum of 6 cubed plus 6 squared; that is, $6^3 + 6^2 = 252$.The fact $12x = 6$ gives $x = \frac{1}{2}$, which is the length. The width comes to be $\frac{2}{3} \times \frac{1}{2} = \frac{1}{3}$.

Note: We mentioned previously that the Mesopotamian mathematicians transformed equations of the form

$x^3 + bx^2 = c$ to the form $\frac{x^3}{b^3} + \frac{x^2}{b^2} = \frac{c}{b^3}$ by multiplying by $\frac{1}{b^3}$. Setting

$n = \frac{x}{b}$, we get the table format $n^3 + n^2 = d$, where d is known.

For further reading on tablet BM 85200 + VAT 6599, see Friberg (2007, 58, 61, 62) and Hoyrup (2002, 137–49).

CHAPTER 8 Problem Texts: Algebraic-Geometric Exercises

During the excavations (1945–61) at Tell Harmal and Tell Dhibai, the renowned Iraqi archeologist Taha Baqir had discovered a significant collection of cuneiform tablets dealing with legal issues and with mathematical problems. Among them was the code of laws of Eshnunna, which precedes the famous code of Hammurabi by a century. Twelve important mathematical cuneiform tablets (dating back to the early second millennium BCE) were deciphered and published by Taha Baqir in Sumer (1950a, 1950b, 1951, and 1962). Those tablets bear the identification numbers: IM 52301, IM 53953, IM 53957, IM 53961, IM 53965, IM 54010, IM 54011, IM 54464, IM 54478, IM 54538, IM 54559, and IM 55357. A more recent discussion of these twelve tablets appears in Goncalves (2015). Tablet IM 55357 has extraordinary significance because it deals with the principles of similar triangles, such as proportionality. Figure 2 is a hand-written copy of tablet IM 55357 by Baqir (1950a). The figure shows at its top a drawing of a right triangle with numbers indicating values of the sides and the areas of the smaller triangles within the main triangle. The tablet shows accurate calculations of the areas of the smaller triangles. The text and its worked-out solution appear below the drawing. Figure 3 reproduces the drawing with numbers written in the sexagesimal system and some letters inserted for identification. In what follows, we give a translation of tablet IM 55357 followed by a present-day mathematical interpretation.

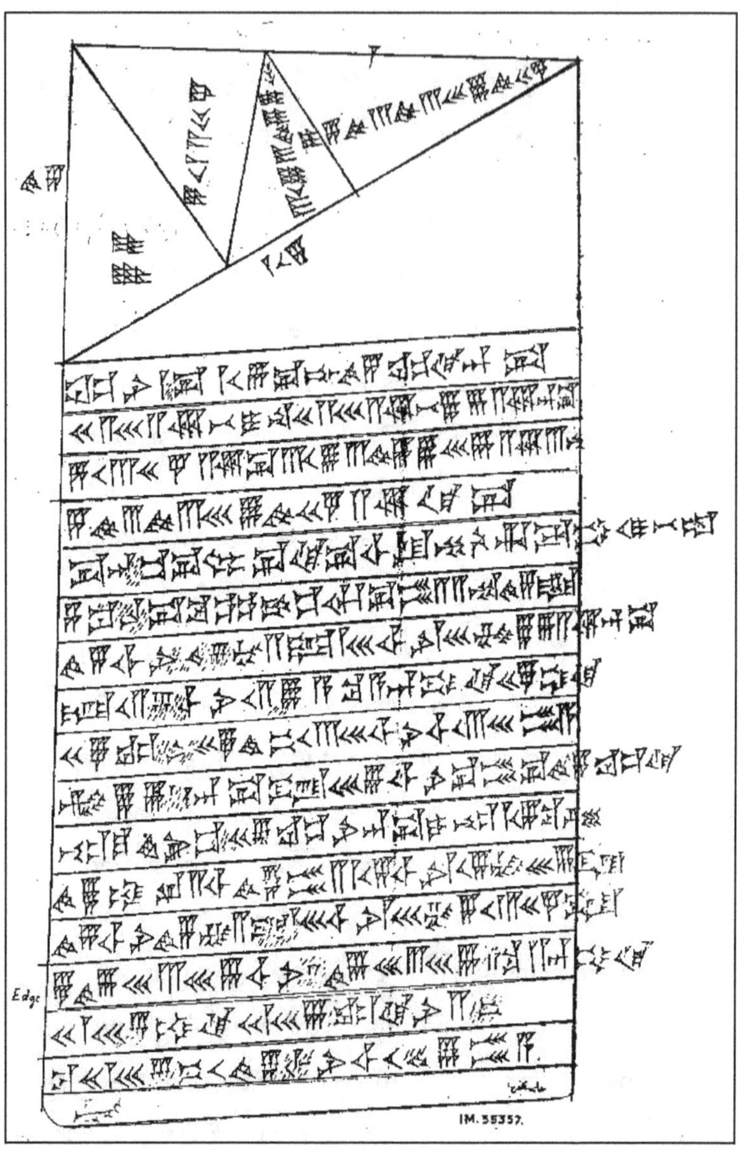

Figure 2. Taha Baqir's hand-copy of tablet IM 55357

Tablet IM 55357 (1800 BCE)

Translation of the given data part: tablet IM 55357

A (right) triangle has sides 0,45 (=45 in decimal), 1,0 (=60), and a hypotenuse of 1,5 (=75). The area, 22,30 (=1350), of the main triangle is cut off into four smaller right triangles by first constructing a perpendicular from the right angle of the main triangle onto its hypotenuse. This process is repeated by constructing perpendiculars onto the hypotenuses of the smaller triangles. The areas of the four smaller triangles are inscribed on the tablet as:

8,6 (=486) for ABD, 5,11;2,24 (=311.04) for ADE, 3,19;3,56,9,36 (=199.0656) for EDF, and 5,53;53,39,50,24 (=353.8944) for EFC.

See figure 3.

Translation of the question part: tablet IM 55357

It was required to calculate the lengths of the sides of triangle ABD and those of all other smaller triangles.

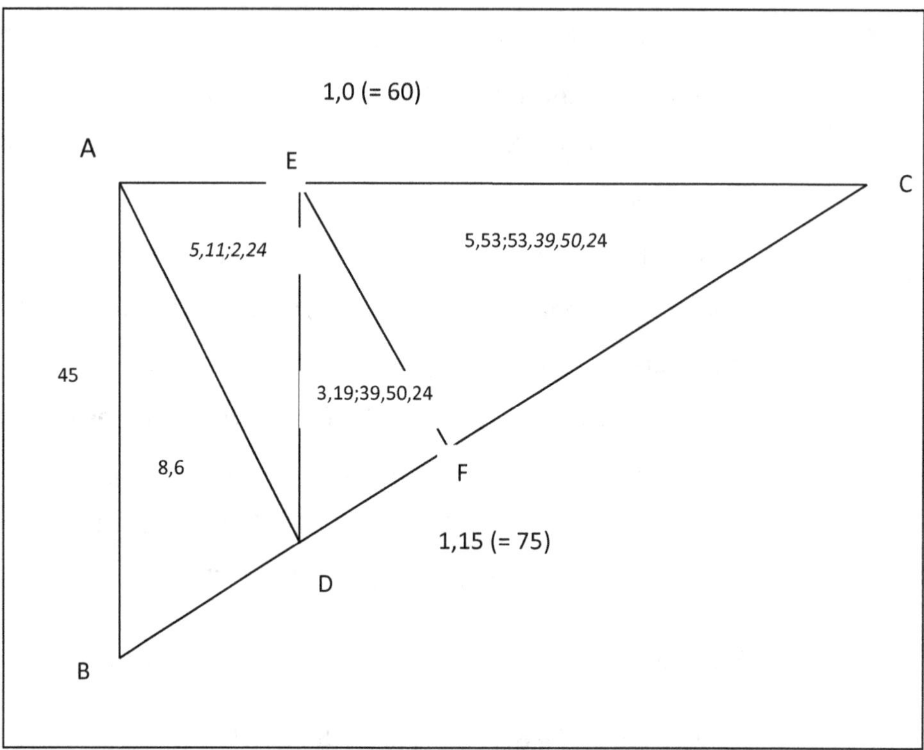

Figure 3. Illustration of Tablet IM 55357

Translation of the solution part: tablet IM 55357

Unfortunately, only the calculations for sides BD and AD of triangle ABD were preserved because the tablet is broken in the middle of the calculations for DE.

Finding BD: Divide 45 by 60 to come up with $\frac{3}{4}$. Multiply $\frac{3}{4}$ by 2 to obtain $1\frac{1}{2}$. Multiply 486, the given area of ABD, by $1\frac{1}{2}$ to obtain 729. Take the square root of 729 to obtain 27. That is the length of BD.

The following formula summarizes the steps for finding BD:

$$BD = \sqrt{\frac{AB}{AC} \times 2 \times \text{area of } ABD} = \sqrt{\frac{45}{60} \times 2 \times 486} = 27$$

Finding AD: Take half of BD. Divide 486, the area of ABD, by half of BD, to get 36. That is the length of AD. In summary,

$$AD = area\,of\ ABD \div \frac{1}{2}BD = 486 \div \frac{1}{2}(27) = 36.$$

Present-day mathematical interpretation: tablet IM 55357

Calculating side BD of triangle ABD:

Baqir (1950a) asserts that the Mesopotamians employed the properties of similar triangles in their solution. Specifically, they applied Euclid's Proposition 8, Book VI which states:

> If in a right-angled triangle a perpendicular is drawn from the right angle to the base, the triangles adjoining the perpendicular are similar both to the whole and to one another.

Further, the Mesopotamians applied the property that corresponding sides of similar triangles are proportional. Since triangles ABD and ADC are similar, then their sides are proportional; that is $\frac{AB}{AC} = \frac{BD}{AD}$.

Therefore,

$$\frac{BD}{AD} = \frac{45}{60} \qquad (8.1)$$

Thus, the area of ABD is

$$(BD \times AD)/2 = 486 \qquad (8.2)$$

Multiplying Eq. (8.1) by Eq. (8.2) and multiplying the result by 2, we obtain

$$\frac{BD}{AD} \times \frac{BD \times AD}{2} \times 2 = \frac{45}{60} \times 486 \times 2$$

$$BD^2 = \frac{45}{60} \times 486 \times 2 = 729$$

$$BD = \sqrt{\frac{45}{60} \times 486 \times 2} = \sqrt{729} = 27$$

Calculating side AD of triangle ABD:

After knowing the value of *BD*, the value of AD is easily calculated as follows:

$$area\,of\ ABD = AD \times \frac{1}{2}BD$$

$$AD = \text{area of } ABD \div \frac{1}{2}BD$$

$$= 486 \div \frac{1}{2}(27) = 36.$$

Note 1: The values 45, 60, and 75 of the sides of the main triangle, ABC, constitute a Pythagorean triple in the ratio: 3: 4: 5 and so do the sides of the other smaller triangles.

Note 2: The areas inscribed inside each smaller triangle were calculated to an incredible accuracy of finite sexagesimal fractions. For example, the area of triangle EFC was given as:

$$5,53;53,39,50,24 = 5 \times 60 + 53 + \frac{53}{60} + \frac{39}{60^2} + \frac{50}{60^3} + \frac{24}{60^4}$$
$$= 353.8944 \text{ in decimal representation.}$$

We discuss now two of several interesting ancient mathematical problems concerning sliding a stick laid against a wall. The Mesopotamians solved those problems by using the Pythagorean rule.

Tablet BM 85196 # 9

A stick (pole, reed, or rod) of length 30 units stands vertically against a wall. The stick upper end has descended a distance of 6 units, how far has the lower end moved? Subtract 6 from 30 to come up with 24. Square 30 to get 900 (= 15,0 sexagesimal). Square 24 to get 576 (= 9,36). Subtract 576 from 900 to come up with 324 (= 5,24). The square root of 324 is 18. Thus the procedure.

Present-day symbolic mathematics: tablet BM 85196 # 9

Examining figure 4, we find:

$b = c - 6 = 30 - 6 = 24$

$a = \sqrt{c^2 - b^2}$, by the Pythagorean rule

$a = \sqrt{30^2 - 24^2} = \sqrt{900 - 576} = \sqrt{324} = 18.$

For further reading on tablet BM 85196, see Hoyrup (2002, 275).

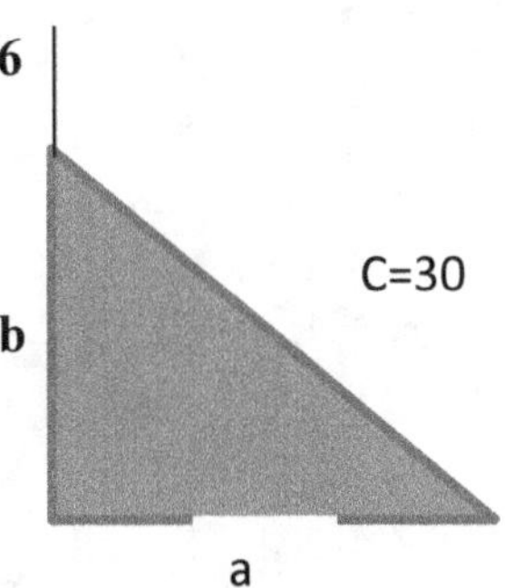

Figure 4. Illustration of Tablet BM 85196 # 9

Tablet BM 34568 # 12

A stick stands vertically against a wall. When the stick upper end descends a distance of 3 units, the lower end moves 9 units. What are the length of the stick and the height of the wall?

In present-day symbolic mathematics, the Mesopotamian solution proceeded as follows (see figure 5):

It is given that $b = c - 3 \ and \ a = 9.$

The Pythagorean rule leads to :

$$c^2 = b^2 + a^2$$

$$c^2 = (c-3)^2 + 9^2$$

$$c^2 = c^2 - 6c + 3^2 + 9^2$$

$$6c = (3^2 + 9^2)$$

$$c = \frac{1}{6}(9^2 + 3^3) = 15.$$

Therefore,

$$b = \sqrt{c^2 - a^2} = \sqrt{15^2 - 9^2} = \sqrt{144} = 12.$$

See Hoyrup (2002, 391–94).

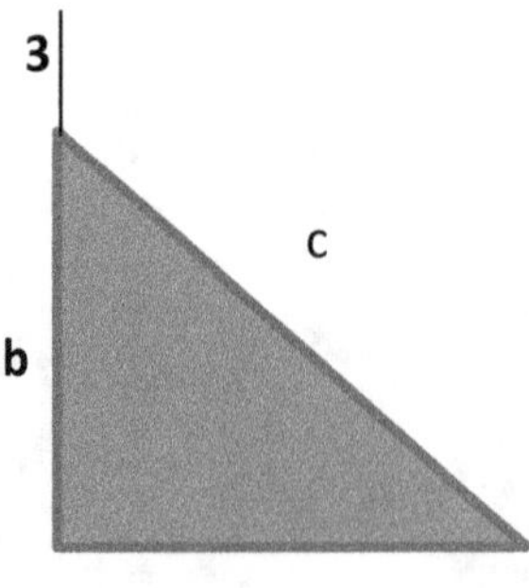

Figure 5. Illustration of Tablet BM 34568 # 12

Tablet Db_2-IM 67118

Tablet IM 67118 was discovered during the excavations of the Iraqi Directorate of Antiquities in 1961 at Tell Dhebai near Baghdad. We present Baqir's (1962) translation and interpretation of the tablet. The reader may also refer to Friberg (2007, 205).

Translation: tablet Db_2-IM 67118

If asked about a rectangle with diagonal $1\frac{1}{4}$ (= 1;15 sexagesimal) and area $\frac{3}{4}$ (= 0;45), what are the length and width? You procede to multiply the diagonal $1\frac{1}{4}$ by itself. You get $1\frac{9}{16}$ (= 1;33,45); hold it. Multiply the area $\frac{3}{4}$ by 2 to get $1\frac{1}{2}$. Subtract this result from $1\frac{9}{16}$ to get a remainder of $\frac{1}{16}$ (= 0;3,45). Take the square root of $\frac{1}{16}$ to get $\frac{1}{4}$ and take half of that to come up with $\frac{1}{8}$ (= 0;7,30). Multiply $\frac{1}{8}$ by $\frac{1}{8}$ to get $\frac{1}{64}$ (= 0;0,56,15), Add the area $\frac{3}{4}$ to $\frac{1}{64}$ to get $\frac{49}{64}$ (= 0;45,56,15). Take the square root of $\frac{49}{64}$ to get $\frac{7}{8}$ (= 0;52,30). Hold another $\frac{7}{8}$. Add the $\frac{1}{8}$, which you squared, to $\frac{7}{8}$. You come up with 1, which is the length of the rectangle. Subtract $\frac{1}{8}$ from the other $\frac{7}{8}$. You obtain $\frac{3}{4}$, which is the width of the rectangle.

Moreover, the ancient mathematicians verified the solution by asking the reverse question: If a rectangle has length 1 and width $\frac{3}{4}$, what are the area and the diagonal? Square the length 1 to get 1 and square the width $\frac{3}{4}$ to get $\frac{9}{16}$. Add this to the length square 1 to obtain $1\frac{9}{16}$. Take the square root of $1\frac{9}{16}$ to get $1\frac{3}{4}$, which is the diagonal. Multiply the length by the width to get $\frac{3}{4}$, which is the area of the rectangle. Thus the procedure.

Present-day symbolic mathematics (Baqir 1962): tablet Db_2-IM 67118

Tablet Db_2-IM 67118 deals with finding the sides of a rectangle having a diagonal of $1\frac{1}{4}$ and an area of $\frac{3}{4}$. Denoting the length and width of the rectangle by x and y, we get the first equation: $xy = \frac{3}{4}$. The Mesopotamian mathematicians squared the diagonal $1\frac{1}{4}$ to obtain $1\frac{9}{16}$. The second equation is $x^2 + y^2 = 1\frac{9}{16}$, which is another demonstration of the use of the Pythagorean theorem. Multiply the area by two to get

$$2xy = 2 \times \frac{3}{4} \quad \Rightarrow 2xy = 1\frac{1}{2}.$$

Subtract twice the area from the square of the diagonal

$$x^2 + y^2 - 2xy = 1\frac{9}{16} - 1\frac{1}{2}$$
$$x^2 + y^2 - 2xy = \frac{1}{16}.$$

Then the Mesopotamian mathematician applied the expansion rule:

$$(a-b)^2 = a^2 + b^2 - 2ab$$

Thus, $(x-y)^2 = \frac{1}{16}$. Taking square roots leads to: $x - y = \frac{1}{4}$.

Let us pause here and wonder why, after having the two equations $x - y = \frac{1}{4}$ and $xy = \frac{3}{4}$, didn't the ancient mathematician substitute $y = x - \frac{1}{4}$ into $xy = \frac{3}{4}$ to obtain $x\left(x - \frac{1}{4}\right) = \frac{3}{4}$? In Baqir's (1962) opinion, by not substituting, the Mesopotamian mathematician demonstrated a masterful maneuver to avoid creating the second-degree equation $x^2 - \frac{1}{4}x = \frac{3}{4}$. Instead, they invoked the expansion formula for

$$(a-b)^2 = a^2 + b^2 - 2ab.$$

Now we continue the solution. Halving $(x-y) = \frac{1}{4}$, you come up with

$$\frac{x-y}{2} = \frac{1}{8}.$$

Squaring,

$$\left(\frac{x-y}{2}\right)^2 = \frac{1}{64}$$

$$\frac{x^2}{4} + \frac{y^2}{4} - \frac{1}{2}xy = \frac{1}{64}.$$

Add the area $xy = \frac{3}{4}$ to the last equation, you get

$$\frac{x^2}{4}+\frac{y^2}{4}-\frac{1}{2}xy+xy=\frac{1}{64}+\frac{3}{4}$$
$$\frac{x^2}{4}+\frac{y^2}{4}+\frac{1}{2}xy=\frac{49}{64}$$
$$\left(\frac{x+y}{2}\right)^2=\frac{49}{64}$$
$$\frac{x+y}{2}=\frac{7}{8}.$$

Add $\frac{x-y}{2}=\frac{1}{8}$ to equation $\frac{x+y}{2}=\frac{7}{8}$ to come up with $x=1$, which is the length of the rectangle.

Subtract $\frac{x-y}{2}=\frac{1}{8}$ from equation $\frac{x+y}{2}=\frac{7}{8}$ to get $y=\frac{3}{4}$. That is the width of the rectangle.

Tablets IM 52916 and IM 52685

Tablets IM 52916 and IM 52685 are two examples of the Tell Harmal collection that contain classified listing of many algebraic equations including quadratics; see Goetz (1951). Both tablets date to the early second millennium BCE. A fragment of a third tablet was also found and registered as IM 52304. Although these tablets are incomplete, their preserved parts contain an elaborate listing of quadratic equations classified according to the type of their solutions. Such tabulated listings provide evidence that the Mesopotamian mathematics was based on some laws and rules. Assuming that m,n are known numbers, some of the classified second-degree equations were:

$$x^2 + nx = b,$$

$$x^2 + x + \frac{x}{2} = b,$$

$$x^2 + mx + \frac{x}{n} = b,$$

$$x^2 - xm = b,$$

$$x^2 = \frac{x}{n} - b.$$

Another category deals with adding and/or subtracting areas of squares to or from each other such as the following:

$$x^2 + y^2 = a,$$

$$x^2 + y^2 + z^2 = a,$$

$$x^2 - y^2 = b.$$

We discuss now two cuneiform tablets that demonstrate the skills of the Mesopotamians in using our present-day "**auxiliary variables**" or "**parameters**" method.

Tablets IM 52916 and IM 52685

Translation:

The excess of the length over the width of a rectangle was squared and subtracted from its area to come up with 500 (= 8,20 sexagesimal). If the length is 30 (= 30), what is the width of the rectangle? Square 30 to get 900 (= 15,0). Subtract 500 from 900 to get 400 (= 6,40). Take half of the length 30 to get 15 (= 15). Square the 15 to get 225 (= 3,45). Add 225 to 400 to get 625 (= 10,25). Take the square root of 625 to get 25. Subtract 15 from 25 to get 10. Subtract 10 from 30 to come up with 20. That is the width of the rectangle.

Present-day symbolic mathematics: tablets IM 52916 and IM 52685

If x and y denote the length and width of the rectangle, the ancient text states that:

$$x = 30, \quad xy - (x-y)^2 = 500$$

Sustitute the identity $y = 30 - x + y$ in $xy - (x-y)^2 = 500$ to get

$$30(30 - x + y) - (x-y)^2 = 500$$
$$900 - 30(x-y) - (x-y)^2 = 500$$
$$(x-y)^2 + 30(x-y) = 400.$$

At this point, the Mesopotamian mathematician invoked the use of an **auxiliary variable** by denoting $(x - y) = z$, which leads to the quadratic equation $z^2 + 30z = 400$. Then, the ancient text applied the method of completing the square to solve this quadratic equation as follows:

$$z^2 + 30z = 400$$
$$\frac{1}{2}(30) = 15 \qquad \text{[halving]}$$
$$15 \times 15 = 225 \qquad \text{[squaring]}$$
$$z^2 + 30z + 225 = 400 + 225 \qquad \text{[completing]}$$
$$(z+15)^2 = 625 \qquad \text{[completing]}$$
$$z + 15 = \sqrt{625} = 25 \qquad \text{[extracting]}$$
$$z = 25 - 15 \qquad \text{[adjusting]}$$
$$z = 10.$$

Then ,

$$x - y = 10$$
$$30 - y = 10$$

$y = 20$; that is the width.

Below, we discuss tablet AO 8862 # 1 that provides another demonstration of the algebraic flavor of the Mesopotamian mathematics. You are asked to find the sides of a rectangle if the sum of its area with the difference of its two sides is 183, and the sum of its two sides is 27. The algebraic setup for this case is $xy+(x-y)=183\,and\;x+y=27$

Tablet AO 8862 # 1

Translation:

I have multiplied the length by width to obtain the area. Then I added to the area, the excess of the length over the width to come up with 183 (= 3,3). I added the length to the width to come up with 27. How much are the length, width, and the area?

Your solution to the problem starts by adding 27, the sum of length and width, to 183 to come up with 210 (= 3,30). Add 2 to 27 to come up with 29. Halve 29 to get $14\frac{1}{2}$. Multiply $14\frac{1}{2}$ by $14\frac{1}{2}$ to get $210\frac{1}{4}$ (= 3,30; 15). Subtract 210 from $210\frac{1}{4}$ to get $\frac{1}{4}$ (= 0;15). Square root of $\frac{1}{4}$ is $\frac{1}{2}$. Add $\frac{1}{2}$ and $14\frac{1}{2}$ to come up with 15; that is the length. Subtract $\frac{1}{2}$ from $14\frac{1}{2}$ to get 14, which is the (false) width. Subtract 2, which you have added to 27, from 14 to get 12. That is the actual width. Multiply the length 15 by the width 12 to come up with the area of 180.

Present-day symbolic mathematics: tablet AO 8862 #1

Denoting the length by x and the width by y, the algebraic form of the problem becomes:

$$xy + x - y = 183$$
$$x + y = 27$$

Adding the two equations gives

$$xy + 2x = 210 \quad \Rightarrow x(y+2) = 210.$$

At the next step, the Mesopotamian mathematician invoked the use of an auxiliary variable (or parameter) by considering $(y+2)$ as one entity, which we denote by m . Thus, the last equation becomes $xm = 210$. Adding 2 to $x + y$, gives $x + (y+2) = 27 + 2$ or $x + m = 29$. Thus, we get the two transformed equations:

$x + m = 29$ and $xm = 210$.

Substituting $m = 29 - x$ into $xm = 210$ gives

$x(29 - x) = 210$ or $x^2 - 29x = 210$.

Then the ancient mathematician solved $x^2 - 29x = 210$ by the method of completing the square, the solution of which is $x = 15$. That is the length. Substituting $x = 15$ into $x + m = 29$ gives $m = 14$. Thus $y + 2 = 14$, which the ancient mathematician referred to as the false width. The true width, therefore, is $y = 14 - 2 = 12$.

Tablets from Susa

In 1936, French archeologists discovered at Susa (capital of Elam in Iran) several cuneiform tablets (dating back to the Old Babylonian Age) that involve mathematical and geometrical cases. One tablet dealt with finding the radius of a circle containing an equilateral triangle of sides 50, 50, and 60. The radius was found to be $31\frac{1}{4}$. See Neugebauer (1969, 46–48).

Some Other Interesting Tablets:

VAT 6598: This tablet further demonstrates the use of the Pythagorean rule by the Mesopotamians. The text asks for the diagonal of a rectangular door, the length and width of which are 40 and 10. For further reading on VAT 6598, see Hoyrup (2002, 267–72).

VAT 7848 #3: The text asks to calculate the area of an isosceles trapezoidal field whose bases are a = 14 and b = 50, and side length s = 30. See figure 6. The solution proceeds as follows:

First, find the height, h, by the Pythagorean rule

$$h = \sqrt{30^2 - 18^2} = 24$$

The area, A, of the trapezoid was then calculated according to:

$$A = \frac{a+b}{2} \times h = \frac{14+50}{2} \times 24 = 768.$$

For further reading on Tablet VAT 7848, see Neugebauer (1945, 141–42).

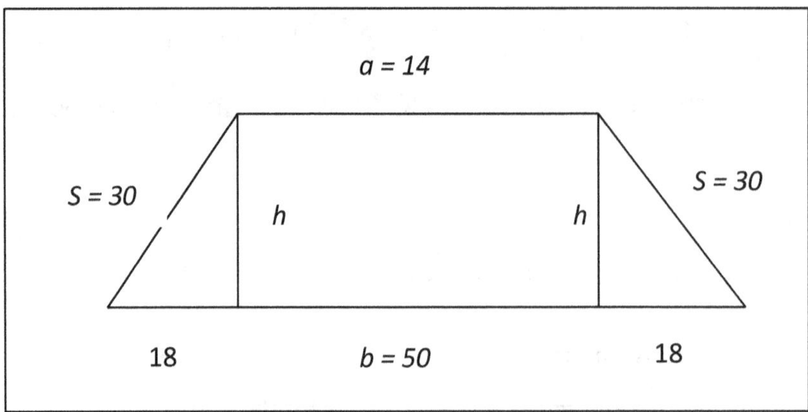

Figure 6. Illustration of Tablet VAT 7848 # 3

Tablet YBC 4675 # 5: A trapezoid has bases *a* and *b*. A line *x* parallel to the bases divided the parallelogram into two equal areas as in figure 7. The Mesopotamians calculated *x* using the formula:

$$x^2 = \frac{1}{2}\left(a^2 + b^2\right)$$

For further reading on YBC 4675, see Neugebauer and Sachs (1945, 44–48).

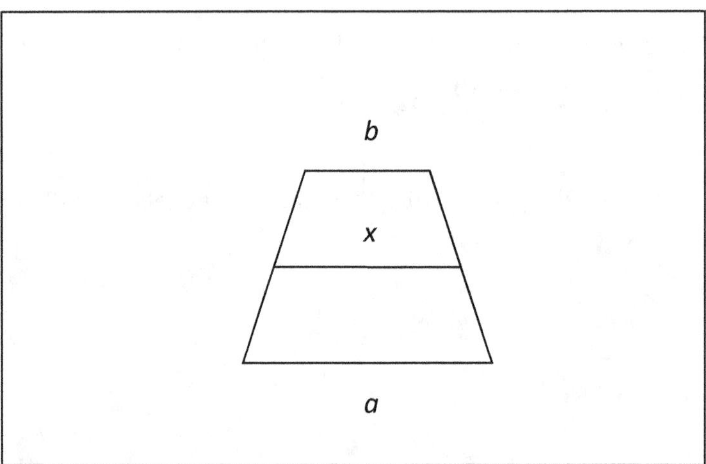

Figure 7. Illustration of Tablet YBC 4675 # 5

Tablet VAT 8512

Note 1: We divide the original text into four blocks: **Block 1** contains the given data. **Block 2** contains calculations for the parallel bar, *x*. **Block 3** contains calculations for the upper height h_U and upper area A_U .

Block 4 contains calculations for the lower height h_L and lower area A_L.

Note 2: We assume that we have a right triangle. Other types of triangles will work too, but that makes the interpretation more complicated. See Figure 8.

Translation: tablet VAT 8512 (figure 8)

[Block 1:] A triangle of base (or width) 30 (= 30) was partitioned into two areas by a bar parallel to its base. The upper area exceeds the lower by 420 (= 7,0 sexagesimal). The lower altitude exceeds the upper altitude by 20 (= 20). What are the two altitudes, the bar, and the two areas?

[Block 2:] Hold the base 30 and hold 420 (= 7,0), which is the excess of the upper area over the lower area. Hold 20 the excess of the lower altitude over the upper altitude. Take the reciprocal of 20 to come up with $\frac{3}{60}$ (= 0;3). Multiply $\frac{3}{60}$ by 420 the excess of the upper area over the lower area, to come up with 21 (= 21).

Hold 21 in your head. Add 21 to the base 30 to come up with 51 (= 51). Multiply 51 by 51 to come up with 2601 (= 43,21). Multiply 21, which you held in your head, by 21 to come up with 441 (= 7,21).

Add 441 to 2601 (= 43,21) to come up with 3042 (= 50,42). Halve 3042 to get 1521 (= 25,21). What is the square root of 1521?

The square root is 39 (= 39). From 39 subtract the number 21 which you squared, to come up with 18. That is the length of the (parallel) bar.

[Block 3:] Well, if the bar was 18, what are the altitudes and the areas of the two parts of the triangle?

Subtract from 51 the number 21, which you squared, to come up with 30 (= 30). Halve 30, and multiply the result 15 by 30 that remained. The result is 450 (= 7,30). Hold it in your head. Multiply the parallel bar 18 by itself to get 324 (= 5,24). Subtract 324 from 450, which you held in your head, to get 126 (=2,6).

What should I do to the 126 to come up with 420, the excess of the upper area over the lower? I multiply $3\frac{1}{3}$ (= 3;20) by 126 to come up with 420.

What is the excess of the base 30 over the parallel line 18? It exceeds it by 12 (= 12). Multiply 12 by $3\frac{1}{3}$, the amount you held in your head, to get 40. That is the length of the upper altitude. Well, if 40 is the upper altitude, what is the upper area?

Add the base 30 to the parallel bar 18 to get 48. Halve 48 (= 48) to get 24 (= 24). Multiply 24 by 40 the upper altitude to get 960 (= 16,0). Therefore, 960 is the upper area.

[Block 4:] Well, if 960 (=16,0) is the upper area, what is the lower height and the lower area? Add the upper altitude 40 to 20, the excess of the lower altitude over the upper; you come up with 60 (= 1,0). That is lower altitude.

Halve the parallel bar 18; you get 9 (= 9). Multiply 9 by 60, the lower altitude, you come with 540 (= 9,0). That is the lower area.

Present-day symbolic mathematics: tablet VAT 8512 (figure 8)

Block 1: The given data are:

Let x be the bar parallel to the base of the triangle.
Let b be the base, A_U and A_L be the upper and lower areas, and h_U and h_L be the upper and lower heights.
It is given that $b = 30, A_U - A_L = 420$, and $h_L - h_U = 20$.
What are the heights, the bar, and the two areas?

See figure 8.

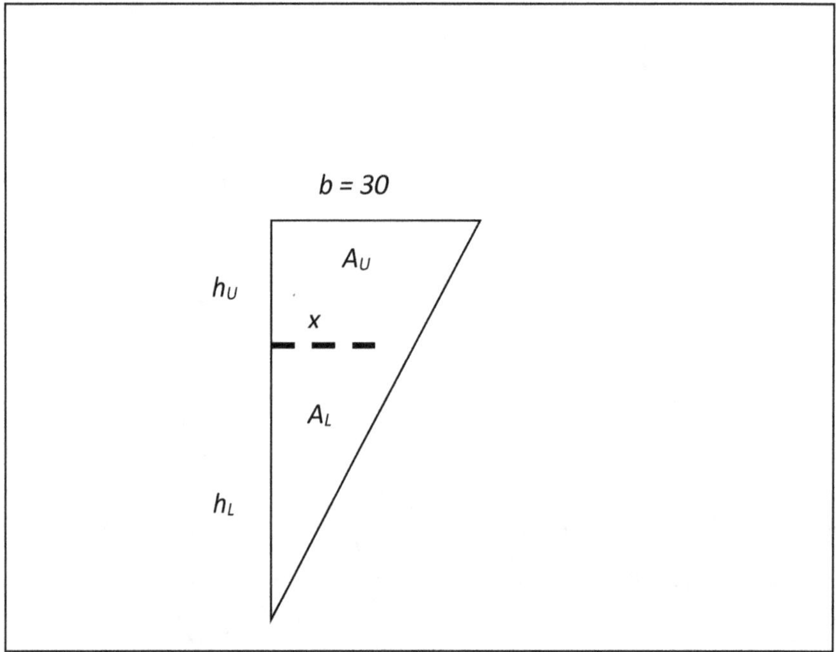

Figure 8. Illustration of Tablet VAT 8512

Block 2 : Finding the parallel bar x

Take reciprocal of $20 = h_L - h_U$

Multiply it by $420 = A_U - A_L$ means : $\dfrac{A_U - A_L}{h_L - h_U} = \dfrac{420}{20} = 21$

Add 21 to 30 means : $\dfrac{A_U - A_L}{h_L - h_U} + b = 51$

Multiply 51 by 51 means : $\left(\dfrac{A_U - A_L}{h_L - h_U} + b\right)^2 = 2601$

Multiply 21 by 21 means : $\left(\dfrac{A_U - A_L}{h_L - h_U}\right)^2 = 441$

Add 2601 to 441 = 3042 means :

$$\left(\frac{A_U - A_L}{h_L - h_U} + b\right)^2 + \left(\frac{A_U - A_L}{h_L - h_U}\right)^2 = 3042$$

Halve 3042 means :

$$\frac{1}{2}\left\{\left(\frac{A_U - A_L}{h_L - h_U} + b\right)^2 + \left(\frac{A_U - A_L}{h_L - h_U}\right)^2\right\} = 1521$$

What is square root of 1521 means :

$$\sqrt{\frac{1}{2}\left\{\left(\frac{A_U - A_L}{h_L - h_U} + b\right)^2 + \left(\frac{A_U - A_L}{h_L - h_U}\right)^2\right\}} = 39$$

From 39 subtract 21 means :

$$\sqrt{\frac{1}{2}\left\{\left(\frac{A_U - A_L}{h_L - h_U} + b\right)^2 + \left(\frac{A_U - A_L}{h_L - h_U}\right)^2\right\}} - \frac{A_U - A_L}{h_L - h_U} = 18 = x;$$

that is the parallel bar, x.

Block 3 : Finding the upper height h_U and the upper area A_U

Take reciprocal of $20 = h_L - h_U$

Subtract from 51 the number 21:

$$\frac{A_U - A_L}{h_L - h_U} + b - \frac{A_U - A_L}{h_L - h_U} = b$$

Halve 30 and multiply it by 30 means : $\frac{1}{2}b^2 = 450$

Muttiply the bar 18 by itself means : $x^2 = 324$

Subtract 324 from 450 means : $\frac{1}{2}b^2 - x^2 = 126$

Add 2601 to 441 = 3042 means :

$$\left(\frac{A_U - A_L}{h_L - h_U} + b\right)^2 + \left(\frac{A_U - A_L}{h_L - h_U}\right)^2 = 3042.$$

What should I do with 126 to come up with 420? Means :

What is the value of k satisfying $k\left(\frac{1}{2}b^2 - x^2\right) = A_U - A_L$?

$$k = \frac{A_U - A_L}{\left(\frac{1}{2}b^2 - x^2\right)} = 3\frac{1}{3}$$

What is the excess of the base over the parallel bar? $(b - x) = (30 - 18) = 12.$

Multiply the excess by $k = 3\frac{1}{3}$ means :

$$(b - x)\frac{A_U - A_L}{\left(\frac{1}{2}b^2 - x^2\right)} = 40 = h_U \text{, the upper height.}$$

The upper area A_U is a trapezoid, the value of which is

$$A_U = \left(\frac{30 + 18}{2}\right) \times 40 = \left(\frac{b + x}{2}\right) \times h_U = 450 \text{, the upper area.}$$

Using the last formula, they calculated the area of a trapezoid by multiplying the height by the average of the two parallel sides. This is exactly what present-day mathematicians do.

Block 4 : Finding the lower height and lower area.
If $h_U = 40$ and $h_L - h_U = 20$, then $h_L = 60$.
The lower area is a triangle with base $x = 18$ and height $h_L = 60$.
Thus, the lower area $A_L = \frac{x}{2} \times h_L = \frac{18}{2} \times 60 = 540$.

For further reading on VAT 8512, see Hoyrup (2002, 234–38), Baqir (2013, 76–79), Huber (1955), and Gandz (1948).

Volumes of Solid Geometric Figures

The content of an important tablet, BM 85194, demonstrates that the Mesopotamian mathematicians calculated the volumes of numerous solid figures such as cones, cubes, cylinders, parallelepipeds, prisms, and pyramids. Also, they worked with the truncated forms of those solid figures. Some of their procedures are listed below:

1. The volumes of the prism and the cylinder were calculated by multiplying the base by the height.
2. The volume of the truncated cone (frustum) was calculated approximately by multiplying the height by the average of the upper and lower bases.
3. The volume of a truncated square pyramid was calculated through an approximate formula and an exact formula.

   Approximate formula: multiply the altitude by the average of the two square areas.

   Exact formula: Let *a*, *b*, and *h* be the side lengths and the height of the truncated square pyramid respectively. They calculated the volume *V* by:

$V = h\left\{\left(\frac{a+b}{2}\right)^2 + \frac{1}{3}\left(\frac{a-b}{2}\right)^2\right\}$, which is equivalent to the simpler formula, $V = \frac{h}{3}\left(a^2 + ab + b^2\right)$.

CHAPTER 9 Mathematics of Compound Interest

Interest-bearing accounts can grow by earning interest according to the simple interest method or the compound interest method. In the simple interest method, the interest is paid on the initial investment (the principal) only, and not on the interest earned over time. In the compound interest method, the interest is paid on the principal and on the interest accrued over time. Thus, compound interest is called "interest on the interest." While compound interest is used in most banking and financial transactions, such as in saving accounts and loan repayments, simple interest is used in some short-term financial transactions.

Although compound interest is currently a driving force in most modern economies, it is by no means a modern invention. Cuneiform text tablets from the Sumerian period (ca. 2600–2350 BCE) and from the Old Babylonian period (ca. 2000–1600 BCE) shows concrete evidence of compound-interest calculations for transactions among people and among city-states. Muroi (2015) asserts that the inscriptions on the Entemena Foundation Cone (2400 BCE) offer a strong evidence that **"compound interest began at Sumer."** According to those inscriptions, Entemena (the Sumerian king of Lagash) lent an amount of barley to the city of Umma. War broke out because Umma refused to repay the loan. The inscriptions on the Cone depicts precise calculations of the maturity value of the loan at 33.3 annual percentage rate. Umma lost the war and was forced to repay the loan at maturity value as calculated by Lagash. I will discuss other ancient examples that demonstrate calculations of the time (doubling time) it takes for a principal to double (or grow to higher folds) under compounded interest.

The doubling time of a principal is still of interest in our modern days. The present-day "rule of 70" states that it will take $\frac{70}{100r}$ years for a principal to double at an annual compound interest rate of *r* (in decimal). Thus, at 5 percent annual interest rate, a principal doubles in

$$\frac{70}{100(0.05)} = \frac{70}{5} = 14 \text{ years.}$$

Table 1 in Bakir (2016) gives more rules for a principal to triple, quadruple, etc.

Preliminaries

For the benefit of the reader, we give the basic formulae of the present-day mathematics of compound interest. Let

P = the initial principal (or present value) that is invested,

r = annual interest rate expressed in decimal form. In percent, 100*r* is called the annual percentage rate (APR).

n = frequency of compounding periods per year. For example, *n* = 1, 2, 4, 12, 52, and 365 correspond to annual, semiannual, quarterly, monthly, weekly, and daily compounding, respectively. Also, *n* = 1/2, 1/3, 1/4, 1/5, and 1/10 correspond to compounding every two years (biannual), every three years (triennial), every four years (quadrennial), every five years (quinquennial), and every ten years (decennial). An ancient example of compounding once every five years is presented in tablet VAT 8528.

Further, let

t = time of investment in years,

F = future value of the principal *P*, and

I = total accrued interest.

The basic compound interest formulae are

$$F = P\left(1+\frac{r}{n}\right)^{nt} \tag{9.1}$$

$$I = F - P, \tag{9.2}$$

For continuous compounding, $(n \to \infty)$, the basic formula becomes

$$F = P\lim_{n\to\infty}\left(1+\frac{r}{n}\right)^{nt} = P\left[\lim_{n\to\infty}\left(1+\frac{r}{n}\right)^{n}\right]^{t} = P[e^{r}]^{t} = Pe^{rt}. \tag{9.3}$$

The simple interest method has two main formulae:

$$F = P(1+rt) \quad \text{and} \quad I = Prt.$$

The m-Fold Exact-Time Rule

The question at hand is: How long does it take a principal to grow m-folds at an annual interest rate of *r* and a compounding frequency of *n* times per year?

Solution: Substitute *F* = *mP* in Formula (9.1) to obtain

$$mP = P\left(1 + \frac{r}{n}\right)^{nt}$$

$$m = \left(1 + \frac{r}{n}\right)^{nt}$$

$$\ln m = nt \ln\left(1 + \frac{r}{n}\right)$$

$$nt = \frac{\ln m}{\ln\left(1+\frac{r}{n}\right)} \quad \text{in compounding periods.}$$

In years, the *m*-fold exact-time compounding formula is

$$t = \frac{\ln m}{n\ln\left(1+\frac{r}{n}\right)} \quad \text{in years.} \qquad (9.4)$$

For continuous compounding, the *m*-fold exact-time formula can be derived from (9.3) as:

$$t = \frac{\ln m}{r} \quad \text{in years.} \qquad (9.5)$$

Note that we can set *m* = 1.5 if we require the time for a principal to grow by 50 percent. Of special interest is when *m* = 2, which leads to the doubling exact-time formula

$$t = \frac{\ln 2}{n \ln\left(1+\frac{r}{n}\right)} \quad \text{in years.} \tag{9.6}$$

For continuous compounding, the doubling exact-time formula becomes

$$t = \frac{\ln 2}{r} \quad \text{in years.} \tag{9.7}$$

The m-Fold Approximate-Time Rule

Instead of using the doubling exact-time in Formula (9.6) that requires logarithmic calculations, present-day financiers use a simple rule (the rule of 70) to approximate the required doubling time under compound interest. The rule of 70 states that at an annual interest rate of *r* (in decimal form), a principal doubles in $\frac{70}{100r}$ years. Thus, at a 7 percent APR, money doubles in $\frac{70}{7} = 10$ years. Some people prefer using 72 instead of 70 because 72 has more divisors than does 70; thus people may use the 72 double-your-money rule ($\frac{72}{100r}$). We will extend the rule of 70 (or 72) to rules that give the approximate time for a principal to grow *m*-folds (triple, quadruple, etc.).

Suppose that a principal *P* is invested at an annual interest rate of *r* (in decimal) compounded at a frequency of $n > 0$. It is required to determine the approximate time for the principal to grow *m*-folds, that is, to become *mP*, where $m \geq 1$. Below, we reproduce Bakir's (2016) derivation of the desired approximate-time formula.

Recall that the *m*-fold exact-time Formula (9.4),

$$t = \frac{\ln m}{\mathrm{n}\ln\left(1+\frac{r}{n}\right)} .$$

Using the Maclaurin's series expansion (see any calculus book), we get

$$\ln\left(1+\frac{r}{n}\right)=(r/n)-\frac{(r/n)^2}{2}+\frac{(r/n)^3}{3}-\cdots \quad ,\text{for}\left|(r/n)\right|<1 \quad .$$

Because (r/n) is small, we can ignore terms involving $(r/n)^2, (r/n)^3$, *etc.* Thus,

$$\ln\left(1+\frac{r}{n}\right)\cong(r/n).$$

Therefore, formula (9.4) becomes

$$t=\frac{\ln m}{n\ln\left(1+\frac{r}{n}\right)}\cong\frac{\ln m}{n(r/n)}.$$

The m - fold approximate time formula becomes

$$t\cong\frac{\ln m}{r}, \text{ in years.} \qquad (9.8)$$

When m = 2, we get the famous approximate double-time formula (the rule of 70) as follows:

$$t\cong\frac{\ln 2}{r}=\frac{0.6931}{r}\text{ , or } t\cong\frac{0.70}{r}, \text{ which leads to}$$

$$t=\frac{70}{100r}, \text{ in years.} \qquad (9.9)$$

Note 1: The m-fold approximate-time in formulae (9.8) and (9.9) are indifferent to the compounding frequency, n.

Note 2: Under continuous compounding, the *m*-fold approximate- and the exact-time formulae are identical:

$$t = \frac{\ln m}{r} \quad \text{in years.}$$

Some Mesopotamian Compound Interest Calculations

We discuss two Mesopotamian artifacts that involve compound interest transactions; specifically doubling-time calculations. Mesopotamians used exponential tables and their inverses (logarithmic tables) to answer compound interest questions. Annual interest rates of 10 percent up to 33 percent on loans were common in Mesopotamia in early periods. Later, the code of Law of Hammurabi legislated and lowered those rates to 20 percent.

The first artifact that we discuss is tablet AO 6770 that calculates the time for a principal to double at 20 percent APR compounded annually. The second artifact is tablet VAT 8528 that calculates the time for a principal to grow 64 folds when the frequency of compounding occurs once every five years (quinquennial).

Tablet AO 6770 (ca. 2000 BCE)

In effect, tablet AO 6770 asks: How long does it take a principal to double at 20 percent APR compounded annually?

We will discuss the present-day rule of 70, and the Mesopotamian solutions to the above question, in that order.

Present-day exact solution: tablet AO 6770

Substitute *r* = 20 percent = 0.20 and *n* = 1 in the exact doubling-time formula in (9.6) to obtain

$$t = \frac{\ln 2}{n \ln\left(1 + \frac{r}{n}\right)} = \frac{\ln 2}{\ln\left(1 + \frac{0.20}{1}\right)} = \frac{\ln 2}{\ln(1.20)} = \frac{0.6931}{0.1823} = 3.80 \text{ years}$$

Thus, the present-day exact answer is 3.80 years, which is equivalent to 3 years, 9 months, and 18 days.

Rule of 70 approximate solution: tablet AO 6770:

Substitute *r* = 0.20 in the rule of 70, Formula (9.9), to obtain

$$t = \frac{70}{100r} = \frac{70}{20} = 3.5 \text{ years}$$

Thus, the rule of 70 answer is 3 years and 6 months.

The Mesopotamian solution: tablet AO 6770

The Mesopotamian solution in tablet AO 6770 reduces the question to that of finding the time *t* in the exponential equation $2 = \left(\frac{6}{5}\right)^t$. This is equivalent to our present-day formula $F = P\left(1 + \frac{r}{n}\right)^{nt}$, in which

$F = 2P, n = 1, and\ r = 0.20 = \frac{1}{5}$. As evidenced by excavations, the Mesopotamians had calculated tables of powers of various numbers. By searching through the power tables of $\left(\frac{6}{5}\right)$, they found that $2 = \left(\frac{6}{5}\right)^t$ leads to values of *t* falling between 3 and 4. Through interpolation between

t = 3 and t = 4, their answer was 3 years and $9\frac{4}{9}$ months. This Mesopotamian answer is equivalent to 3 years, 9 months, 13 days and one-third of a day.

It can be seen that the Mesopotamian answer of 3 years, 9 months, 13 days and one-third of a day is very close the present-day exact answer of 3 years, 9 months and 18 days. For further reading on tablet AO 6770, see Curtis (1978) and Muroi (1987).

Tablet VAT 8528 (ca. 2000 BCE)

Tablet VAT 8528 deals with finding the time required for a principal invested at 20 APR to grow 64 folds when compounding occurs once every five years. It asks:

If you lend one mina of silver at an annual interest rate of 12/60 of a mina per year, how long does it take to be repaid as 64 minas?

It had been the custom in Mesopotamia to begin capitalizing the interest when the outstanding principal doubles. In the context of VAT 8528, the annual interest rate is given as $\frac{12}{60} = 20$ percent. Recall the basic simple interest formula, $F = P(1 + rt)$. Substitute $F = 2P, r = 0.20$ to obtain $2 = (1 + 0.20t)$. Solving, we see that doubling the standing principal occurs every $t = 5$ years.

Present-day exact solution: tablet VAT 8528

Accounting for the fact that compounding occurs once every five years, the question in tablet VAT 8528 becomes: How long will it take a principal to grow 64 folds if compounding occurs once every five years at a 20 percent APR?

Substituting m = 64, r = 0.20, n = 1/5 in Formula (9.4), we obtain

$$t = \frac{\ln m}{\mathrm{n}\ln\left(1+\frac{r}{n}\right)} = \frac{\ln 64}{\frac{1}{5}\ln\left(1+\frac{0.20}{1/5}\right)} = \frac{5\ln 64}{\ln 2} = \frac{5(4.1588)}{0.6931} = 30 \text{ years.}$$

The m-fold approximate-time rule: tablet VAT 8528

Substitute m = 64 and r = 0.20 in Formula (9.4), the m-fold approximate-time rule, to obtain

$$t \cong \frac{\ln m}{r} = \frac{\ln 64}{0.20} = \frac{4.1588}{0.20} = 20.8 \text{ years}$$

This approximation of 20.8 years is very far off the exact answer of 30 years. The reason for this gross discrepancy is that the approximation $\ln\left(1+\frac{r}{n}\right) \cong \frac{r}{n}$ requires $\frac{r}{n} < 1$. In our case: $\frac{r}{n} = \frac{0.20}{1/5} = 1.0$,

and $\ln\left(1+\frac{0.20}{1/5}\right) = \ln 2 = 0.6931$ are quite different. The time-approximation rules may become invalid when the compounding frequency is less than once year.

The Mesopotamian solution: tablet VAT 8528

The Mesopotamian solution in tablet VAT 8528 involves reducing the question to that of solving the exponential equation $2^{\frac{t}{5}} = 64$. One can arrive at this equation by substituting

$F = 64, P = 1, r = 0.2, and\ n = \frac{1}{5}$ in Formula (9.1) to obtain

$$F = P\left(1 + \frac{r}{n}\right)^{nt}$$

$$64 = \left(1 + \frac{0.2}{1/5}\right)^{\frac{t}{5}}$$

$$64 = (2)^{\frac{t}{5}}\ .$$

Then the Mesopotamians searched their power tables to look up the power of 2 such that $2^{\frac{t}{5}} = 64$. Knowing that $2^2 = 4$, $2^3 = 8$, $2^4 = 16$, $2^5 = 32$, and $2^6 = 64$, leads to $\frac{t}{5} = 6$. Thus the solution, t = 30 years. Again, the Mesopotamian solution of 30 years coincides with the present-day exact time solution. For further details on tablet VAT 8528 (and a similar tablet, VAT 8521), see Neugebauer (1935–38) and Muroi (1990).

References

Bakir, S. T. 2016. "Compound Interest Doubling Time Rule: Extensions and Examples from Antiquities." *Communications in Mathematical Finance* 5 (2): 1–11. http://www.scienpress.com/download.asp?ID=1875

Baqir, T. 1950a. "An Important Mathematical Problem Text from Tell Harmal (on a Euclidean theorem)." *Sumer* 6: 39–54.

Baqir, T. 1950b. "Another Important Mathematical Text from Tell Harmal." *Sumer* 6: 130–48.

Baqir, T. 1951. "Some More Mathematical Texts from Tell Harmal." *Sumer* 7: 28–45.

Baqir, T. 1962. "Tell Dhibai: New Mathematical Texts." *Sumer* 18: 11–14.

Baqir, T. 1973. *An Introduction to History of Ancient Civilizations* (written in Arabic). Baghdad, Iraq.

Baqir, T. 2013. *A Brief History of the Sciences and Knowledge in the Ancient and Arabic-Islamic Civilizations* (written in Arabic). Createspace, USA.

Bell, E. T. 1945. *Development of Mathematics*. 2nd ed. New York: McGraw-Hill.

Bruins, E. M. 1949. "On Plimpton 322, Pythagorean Numbers in Babylonian Mathematics." *Konninklijke Nederlandse Akademie van Wetenschappen Proceedings* 52: 629–32.

Bruins, E. M. 1955. "Pythagorean Triads in Babylonian Mathematics: the errors on Plimpton 322." *Sumer* 11: 117–21.

Curtis, L. J. 1978. "Concepts of the Exponential Law Prior to 1900." *American Journal of Physics* 46 (9): 896–06.

Friberg, J. 2007. *A Remarkable Collection of Babylonian Mathematical Texts.* New York: Springer.

Gandz, S. 1948. "Studies in Babylonian Mathematics II. Indeterminate Analysis in Babylonian mathematics." *Osiris* 8: 12–14.

Goetze, A. 1951. "A Mathematical Compendium from Tell Harmal." *Sumer* 7: 126-154.

Goncalves, C. 2015. *Mathematical Tablets from Tell Harmal*. New York: Springer.

Hoyrup, J. 2002. *Lengths, Widths, Surfaces: A Portrait of Old Babylonian Algebra and Its Kin.* New York: Springer.

Huber, P. 1955. "Zu einem mathematischen Keilschrifttext (VAT 8512)." *Isis* 46: 104–6.

Kohn, E. 2001. *CliffsNotes Quick Review Geometry*. Hoboken, NJ: Wiley Publishing.

Kohn, E. and Herzog, D. A. 2011. *CliffsNotes Algebra II Quick Review*, 2nd Ed. Hoboken, NJ: Wiley Publishing.

Merzbach, U. C. and Boyer, C. B. 2011. *A History of Mathematics,* 3rd Ed. New York: John Wiley and Sons.

Muroi, K. 1987. "A New Interpretation of Babylonian Mathematical Text AO 6770 (in Japanese)." *Kagakusi Kenkyu* 162: 261–66.

Muroi, K. 1990. "Interest Calculation in Babylonian Mathematics: New Interpretations of VAT 8521 and VAT 8528." *Historia Scientiarum* 39: 29–34.

Muroi, K. 2015. "The Oldest Example of Compound Interest in Sumer: Seventh Power of Four-thirds." arXiv:1510.00330.

Neugebauer, O. (1929). "Zur Geschichte der babylonischen Mathematik." Quellen und Studien zur Geschichte der Mathematik, Astronomie und Physik. Abteilung B: Studien 1 (1929-31), 67-80.

Neugebauer, O. 1935–37. *Mathematische Keilschrift-Texte,* 3 volumes. Berlin: Springer.

Nuegebauer, O. 1969. *The Exact Sciences in Antiquity,* 2nd ed. New York: Dover Publications.

Neugebauer, O. and Sachs, A. 1945. "Mathematical Cuneiform Texts," *American Oriental Series 29*, New Haven, CT: American Oriental Society.

Robson, E. 2008. *Mathematics in Ancient Iraq: A Social History*. Princeton, NJ: Princeton University Press.

Roux, G. 1992. *Ancient Iraq.* New York: Penguin.

Saggs, H. W. F. 1962. *The Greatness That Was Babylon*. New York: Hawthorn Books.

Shekoury, R. N. 2010. *Mesopotamians: Pioneers of Mathematics.* CreateSpace, USA.

Thureau-Dargin. 1938. *Textes Mathematiques Babyloniens*. Liden: Brill.

Index

T

U

W

NOTES

NOTES

www.ingramcontent.com/pod-product-compliance
Lightning Source LLC
Chambersburg PA
CBHW081559220526
45468CB00010B/2697
9781974502615